Annual Meeting of the American Electrophoresis Society 2012

(AES)

Topical Conference at the 2012 AIChE Annual Meeting

Pittsburgh, Pennsylvania, USA
28 October - 2 November 2012

ISBN: 978-1-62276-743-4

Printed from e-media with permission by:

Curran Associates, Inc.
57 Morehouse Lane
Red Hook, NY 12571

Some format issues inherent in the e-media version may also appear in this print version.

Printed by Curran Associates, Inc. (2013)

For permission requests, please contact AIChE
at the address below.

AIChE
3 Park Avenue
New York, NY 10016-5991

Phone: (203) 702-7660
Fax: (203) 775-5177

www.aiche.org

Additional copies of this publication are available from:

Curran Associates, Inc.
57 Morehouse Lane
Red Hook, NY 12571 USA
Phone: 845-758-0400
Fax: 845-758-2634
Email: curran@proceedings.com
Web: www.proceedings.com

TABLE OF CONTENTS

(8a) Advantages of Microgradients for Steady State Separations

Mark Hayes, Chemistry and Biochemistry, Arizona State University, Tempe, AZ

Core to all separations science is resolution: causing important components to move in time or space away from each other. Although a description of core processes controlling resolution across all separations systems does not exist, the most successful techniques are considered to be electrophoretic in nature. Early demonstrations of capillary- and microchip- based electrophoresis have shown absolutely extraordinary resolution using high fields showing deuterated vs. hydrogenated and sub-millisecond separations. These results were obtained in systems operating in a linear mode —spreading and diluting a mixture out along a single axis. Our work uncovers a radically new approach enabled by the short length-scale of microdevices which exploits local field gradients. The core to this discovery is setting up punctuated microgradients separated by flat field zones. This uniquely minimizes the effects of diffusion by dramatically increasing the local restoring forces. As a result, not only does the resolution increase by a factor tied to the strength of the local gradients, but can be accomplished this in parallel while increasing concentration of the targets (focusing). This opens up several new venues for exploitation, including dynamic and programmable collection of specific analytes at higher resolution than is available from any traditional technique in an array format. Speeds of separations are also dramatically increased. We describe the underlying theoretical principles and provide examples of flow/electric field and electrokinetic/dielectrophoretic devices in a microfluidic format. The underlying models extend to any force which can be induced into dispersed microgradients in the presence of a static counter field.

(8b) Surfactant Induced Electroosmotic Flow in Microfluidic Capillaries

Glareh Azadi and Anubhav Tripathi, School of Engineering, Brown University, Providence, RI

Control of electroosmotic flow (EOF) in microfluidic devices is essential in applications such as protein/DNA sizing and high throughput drug screening. With the growing popularity of poly(methyl methacrylate) (PMMA) as the substrate for polymeric based microfludics, it is important to understand the effect of surfactants on EOF in these devices. In this article we present an extensive investigation exploring changes in EOF rate induced by sodium dodecyl sulfate (SDS), polyoxyethylene lauryl ether (Brij35) and cetyltrimethylammonium bromide (CTAB) in PMMA microfluidic capillaries. In a standard protein buffer (Tris-Glycine), microfluidic capillaries exhibited a cathodic EOF with measured mobility of $1.54 \pm 0.1(\times 10^{-4}\,cm^2 / V.s)$. In the presence of surfactant, below a critical concentration, EOF was independent of surfactant concentration. At high concentrations of surfactants, the electroosmotic mobility was found to linearly increase/decrease as logarithm of concentration before reaching a constant value. While with SDS the EOF increased by 257%(compared to buffer), it was decreased by 238% with CTAB. In case of Brij35, the electroosmotic mobility was reduced by 70%. Possible mechanisms are proposed explaining the observed changes in EOF and zeta potential values. In a binary surfactant system of SDS/CTAB and SDS/Brij35, addition of oppositely charged CTAB reduced the SDS induced EOF more effectively compared to non-ionic Brij35. Using neutral polymer coatings in combination with SDS resulted in 50% reduction in the electroosmotic mobility with 0.1% Hydroxypropyl Methyl Cellulose(HPMC) while including 2% Poly (N,N-dimethylacrylamide)(PDMA) had no effect. These results will potentially contribute to the development of PMMA based microfluidic devices.

(8c) Bioparticle Capture in a Sawtooth Dielectrophoretic Microchannel

Paul V. Jones and Mark A. Hayes, Chemistry and Biochemistry, Arizona State University, Tempe, AZ

Rapid bioparticle characterization, as a research goal, has fueled a significant and wide-ranging body of work. Much has been accomplished, but new techniques and applications continue to emerge. Innovations in this field may soon enable the development of rapid, on-site bioanalysis devices that improve the availability, accuracy, and scope of clinical diagnosis. Microfluidic electrokinetic approaches provide unique advantages, including short analysis times, microliter sample and reagent volumes, potentially low cost, and portability. The work presented here explores the use of a single, continuous microchannel in which opposing electrokinetic and dielectrophoretic forces create multiple and distinct bioparticle traps. Specifically, it focuses on capture of bioparticles using DC insulator-based dielectrophoresis in a converging, sawtooth-patterned microchannel. The channel design enables localized isolation and concentration of specific particles based on differences in their physical properties. Various targets with disparate characteristics have been captured and concentrated within the device, including human blood cells and mature amyloid protein fibrils. In each case, reproducible capture occurred at specific locales within the sawtooth channel. Differentiable capture of three live *Escherichia coli* serotypes has also been demonstrated, indicating the potential this technique holds for diagnostic applications.

(8d) Electrochemical Aging Effects of Pt Thin Film Electrodes Under Microfluidic Device Conditions

Aytug Gencoglu, Department of Chemical Engineering, Michigan Technological University, Houghton, MI and Adrienne R. Minerick, Michigan Technological University, Houghton, MI

Many of the functions in a microfluidic device, such as sensing, transport and separation are based on electrodes. Noble metals are very commonly used as electrodes in microfluidic applications, and in the microfluidic electrokinetic realm, they are widely regarded as inert materials. However, they are known to undergo various chemical reactions and morphological changes and to be subject to fouling. In microfluidic devices, subtle changes in electrode behavior and appearance have been observed. Surface changes of Pt electrodes were previously studied whereby Pt electrodes utilized in common buffer compositions yielded surface chemical changes including Pt oxidation, Pt dissolution, Pt redeposition, and fouling by solute species. Chemical changes were observed with SEM-EDS and XPS while morphological changes were observed via SEM and optical light microscopy. However, the effect of these changes on electrode performance was not investigated.

In the present work, the electrochemical behavior of thin film Pt electrodes was evaluated through cyclic voltammetry of a 1 mM $K_3Fe(CN)_6$, 0.5 M KCl solution. The cyclic voltammograms illustrate the redox reactions of the model compound $K_3Fe(CN)_6$ in unused Pt electrodes and were compared to the voltammograms obtained with the same devices after aging in various conditions. The conditions comprised a matrix of phosphate buffer solution conductivities (0.01 and 0.10 S/m) and different applied DC current levels, as well as an applied 5 V_{pp}, 1 kHz AC signal. Different devices of identical design were aged under each set of buffer and electric signal conditions. SEM-EDS was performed on the unused and aged electrodes to evaluate surface chemical and morphological changes. Similar tests of Pt wire electrodes showed that after 6 hrs under a 100 μA current in a 0.10 S/m phosphate buffer solution, $K_3Fe(CN)_6$ redox peak heights were significantly decreased. Further, the cyclic voltammograms were different than the control CV experiments in 0.5 M KCl solution and water electrolysis peaks were persistent. These results suggest that aging weakens the electrochemical response of Pt electrodes, but it does not completely passivate Pt surfaces even after long usage times.

C-MEMS Based Electrodes For The Dielectric Characterization Of Microparticles Employing Dielectrophoresis

Abstract ID# 275448

*Víctor H. Pérez-González,[1,2] Vinh Ho,[2] Lawrence Kulinsky,[2] Blanca H. Lapizco-Encinas,[3] Marc Madou,[2,4] Sergio O. Martínez-Chapa,[1] **

1 BioMEMS Research Chair, Tecnológico de Monterrey, Monterrey, NL, México.
2 Mechanical & Aerospace Engineering, University of California, Irvine, Irvine, CA, US.
3 Microscale Bioseparations Laboratory and Department of Chemical Engineering, Tennessee Technological University, Cookeville, TN, US.
4 Ulsan National Institute of Science and Technology, World Class University Program, Ulsan 689-79, South Korea.

**Correspondence author. Email: smart@itesm.mx*

ABSTRACT

Dielectrophoresis (DEP), the force experienced by polarizable particles exposed to a non-uniform electrical field, is an electrokinetic technique that has gained importance in the last two decades within the microfluidic research community. DEP can be employed to perform manipulation, isolation, filtration, focusing, characterization, or concentration of microparticles. The efficiency of dielectrophoresis is directly related to the distribution of the electrical field that can be controlled by the geometric design of the electrodes. When information about the dielectric properties of particles is readily available, mathematical modeling can be used to find the desired frequency window before carrying out the experiment, saving time and resources. The dielectric characterization of microparticles employing dielectrophoresis was covered in several works; nonetheless there are some problems yet to be solved. In particular, the uncontrolled distribution of the field nonuniformity within the device was demonstrated to affect the dielectric properties estimation. The geometry of the electrodes defines the dielectrophoretic response of the microparticles under analysis. Carbon MEMS (C-MEMS) is a microfabrication process in which carbon microstructures are obtained through photolithography and pyrolysis of organic precursors. The C-MEMS process is especially useful in biotechnology applications because of the wide electrochemical stability and biocompatibility of carbon. Using C-MEMS process, three-dimensional (3D) electrodes can be fabricated, allowing for the development of new electrode geometries different from the widely used interdigitated fingers and polynomial electrodes. This work deals with the development of 3D electrode geometries employing the C-MEMS microfabrication process and construction of microfluidic platforms for the dielectric characterization of microparticles employing DEP. COMSOL Multiphysics was employed to analyze the distribution of the electrical field created by the proposed electrode geometry.

INTRODUCTION

Dielectrophoresis (DEP), the force experienced by polarizable particles exposed to a non-uniform electrical field [1], is an electrokinetic technique that has gained importance in the last two decades within the microfluidic research community. DEP can be employed to separate [2, 3], isolate [4], displace [5-7], trap [8-12], filer [13], and characterize [14-17] microparticles.

The efficiency of dielectrophoresis is directly related to the distribution of the electrical field that can be controlled by the geometric design of the electrodes. An AC signal is applied to carry out a dielectrophoretic experiment. In order to optimize the DEP process for particle manipulation (especially for particles with unknown dielectric properties), a frequency sweep is performed to find the frequency range in which the particle can be manipulated effectively. When information about the dielectric properties of particles is readily available, mathematical modeling can be used to find the desired frequency window before carrying out the experiment, saving time and resources. Through a-priori knowledge of the relative permittivity and electric conductivity of a particle and the medium, it is possible to optimize separation, isolation, manipulation or filtration processes based on DEP.

The dielectric characterization of microparticles employing dielectrophoresis was covered in several works; nonetheless there are some problems yet to be solved. In particular, the uncontrolled distribution of the field nonuniformity within the device was demonstrated to affect the dielectric properties estimation. The geometry of the electrodes defines the dielectrophoretic response of the microparticles under analysis. For instance, if planar interdigitated electrodes are employed to induce DEP, the response of the particles will be of different magnitude depending on their spatial position in the microchannel – the field is stronger closer to the tips of electrodes, and weaker towards the center of the channel. This difference in dielectrophoretic response is a source of measurement uncertainty in characterization applications.

Carbon MEMS (C-MEMS) is a microfabrication process in which carbon microstructures are obtained through photolithography and pyrolysis of organic precursors [18]. The C-MEMS process is especially useful in biotechnology applications because of the wide electrochemical stability and biocompatibility of carbon. Using C-MEMS process, three-dimensional (3D) electrodes can be fabricated, allowing for the development of new electrode geometries different from the widely used interdigitated fingers and polynomial electrodes.

This work deals with the development of 3D electrode geometries employing the C-MEMS microfabrication process and construction of microfluidic platforms for the dielectric characterization of microparticles employing DEP. COMSOL Multiphysics was employed to analyze the distribution of the electrical field created by different electrode geometries. Simulation results informed the optimized design of 3D Carbon electrodes. Applications of this work include environmental screening for water contamination, improvements of clean energy production methods, clinical analysis and food safety control, among others.

THEORY OF DIELECTROPHORESIS

Dielectrophoresis [19] is defined as the interaction between an induced electrical dipole and a non uniform electric field. The dielectrophoretic force experienced by a spherical particle is defined as:

$$\mathbf{F}(\omega) = 2\pi\varepsilon_m r^3 \operatorname{Re}[f_{CM}]\nabla E_{rms}^2 \tag{1}$$

where ε_m is the permittivity of the suspending medium, r is the particle radius, and f_{CM} is the Clausius-Mossotti (CM) factor, which is defined as:

$$f_{CM} = \frac{\varepsilon_p^* - \varepsilon_m^*}{\varepsilon_p^* + 2\varepsilon_m^*} = \frac{\sigma_p^* - \sigma_m^*}{\sigma_p^* + 2\sigma_m^*} \tag{2}$$

where $\varepsilon^* = \varepsilon - j\sigma/\omega$ and $\sigma^* = \sigma + j\varepsilon\omega$ refer to the complex permittivity and conductivity respectively. In (2), subscripts p and m refer to particle and medium respectively, and ω is the angular frequency of the applied electric signal [20].

The CM factor provides information about the polarizability of a particle with reference to the suspending medium. Dielectrophoretic force can pull particles towards the zones of highest electric-field-squared gradient (positive CM factor, positive DEP), or it can repel particles from such zones (negative CM factor, negative DEP).

COMPUTATIONAL SIMULATION

In dielectric characterization applications, one of the most important parameters to control is the distribution of the gradient of the squared rms electric field. A geometry in which the non uniformity of the field presents the same profile in every direction is desired. A computational model was built with COMSOL Multiphysics 4.3a © (COMSOL Inc., Burlington, MA). This model allowed for the analysis of the electric field distribution in the zone of dielectrophoretic activity. COMSOL is a partial differential equation solver that uses the Finite Element Method (FEM) approach to solve physical problems for which no closed form analytical solutions exist. Figure 1 shows the 3D design studied in this work.

Two electrodes are shown in Figure 1. The outer electrode has a ring shape, while the inner one is a post located exactly at the center of the ring electrode. The design resembles the shape of a coaxial cable. This geometry is optimal for the purpose of particle characterization, not only because it meets the requirement of controlling the distribution of field non uniformities, but also because given geometry has a closed form analytical solution that can be used to validate computational simulations. The post has a diameter of 35μm. The separation distance between the edge of the post and the inner wall of the ring is of 75μm. The thickness of the ring electrode is 35μm and the height of both electrodes is 70μm.

Figure 1. "Coaxial cable" electrode geometry for dielectrophoretic characterization of microparticles.

COMSOL Multiphysics Electric Current Module was employed to analyze the model. Boundary conditions of electric potential and ground were employed in the electrodes to run a stationary analysis. The geometry was enclosed in a cube with a surface defined as an electric insulator to account for the microfluidic channel walls. These boundary conditions are needed to solve Laplace's equation. The mesh consisted of 119,608 tetrahedral elements.

RESULTS

By solving Laplace's equation through FEM, the electric field distribution in the proposed geometry was obtained. The profile of the evaluated gradient of the squared electric field (rms) is presented in Figure 2. This gradient profile has several advantages over the one obtained for common interdigitated electrode geometries [16]. Interdigitated electrodes have a profile for the field gradient with a maximum at the tips of each finger and a minimum right at the center of each finger. Since the length of the fingers is usually large, compared to the microscope field of view, it becomes difficult to relate a particle (located at a certain position along the microchannel) with the magnitude of the field gradient associated with that position. On the other hand, the coaxial electrode geometry features a region of operation smaller than the typical field of view of the microscope, thus increasing the likelihood of correctly relating the position of the particles with the value of the field gradient at those spatial points. Another important advantage is that this geometry presents a cylindrical symmetry, meaning that movement of the particles towards the central post, or away from it, will only occur in a radial fashion. This will allow for decreasing the standard deviation of the average velocity of the particles, opposed to the scenario with interdigitated electrodes [16]

The magnitude of the field gradient can be increased employing several different approaches. The first approach is to increase the magnitude of the AC electric potential supplied to the electrodes. However, this will come at a price of also increasing the magnitude of another EK phenomenon (electroosmotic flow, EOF), which would influence particle motion, resulting in an increased measurement error in the dielectrophoretic response. The second approach is to decrease the diameter of the central post while keeping the magnitude of the AC applied voltage constant. The only limitation of this approach is the lower feature size limit of the photolithography process employed to pattern the post.

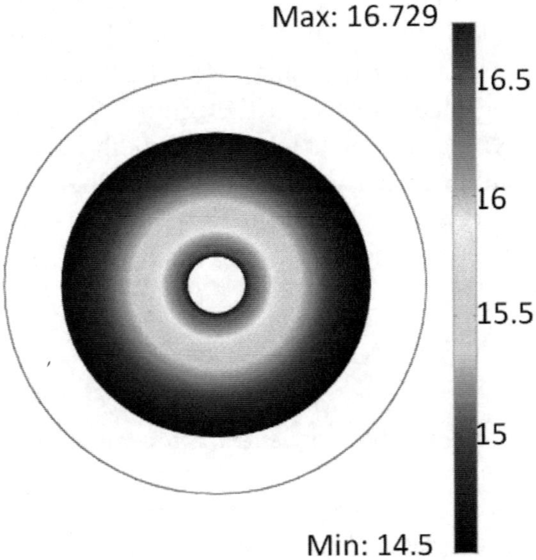

Figure 2. Logarithmic slice plot of the rms electric field squared gradient for the coaxial electrode geometry.

In order to provide a clearer illustration of the features of this design, Figure 3 presents an order of magnitude approximation for the electric field squared gradient. For this, an array of 2x2 dielectrophoretic characterization units was built into COMSOL Multiphysics model. Figure 3a shows such array. Two red lines indicate the segments over which the measurements were taken. Figure 3b shows a line plot of the field gradient over the two segments highlighted in Figure 3a. The profile for this gradient is the same along the two different segments. This feature is an important advantage over the interdigitated electrode geometry, or even over the polynomial electrode geometry [21], commonly used for characterization. In these geometries the field gradient presents different profiles depending on the section over which measurements are taken.

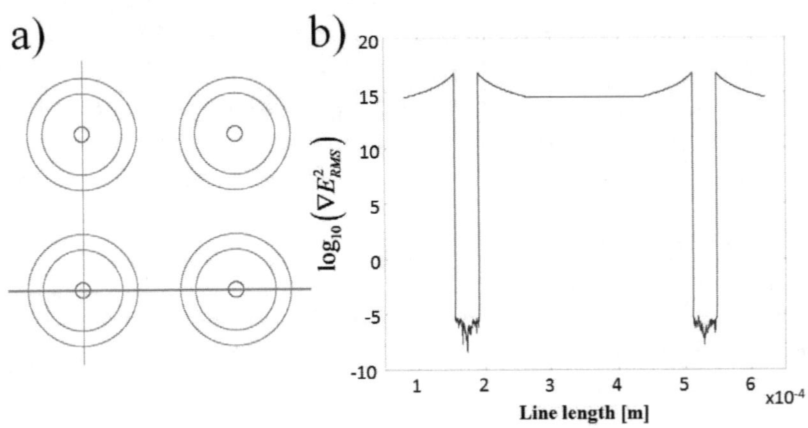

Figure 3. Comparison of the rms electric field squared gradient profile over two different sections of the coaxial electrode geometry. a) Measurement lines, b) Line plot.

CONCLUSIONS

This work has presented a novel geometry for electrodes employed in the dielectric characterization of microparticles using DEP. The coaxial electrode geometry has been shown to present important advantages over conventional electrode designs. Future work includes the fabrication of devices featuring electrodes with the proposed coaxial geometry. Testing will be carried out employing polystyrene microparticles. The methodology presented earlier [16] will be employed to perform the dielectric characterization of the particles. Also, improvement in estimation efficiency will be verified by comparing the results obtained with the coaxial electrode geometry versus those obtained employing interdigitated planar electrodes.

REFERENCES

[1] Pohl, H. A., *Dielectrophoresis : the behavior of neutral matter in nonuniform electric fields*, Cambridge University Press, Cambridge ; New York 1978.

[2] Wang, L., Lu, J., Marchenko, S. A., Monuki, E. S., *et al.*, *Electrophoresis* 2009, *30*, 782–791.

[3] Gallo-Villanueva, R. C., Pérez-González, V. H., Davalos, R. V., Lapizco-Encinas, B. H., *Electrophoresis* 2011, *32*, 2456-2465.

[4] Shafiee, H., Sano, M. B., Henslee, E. A., Caldwell, J. L., Davalos, R. V., *Lab on a Chip* 2010, *10*, 438-445.

[5] Baylon-Cardiel, J. L., Jesús-Pérez, N. M., Chávez-Santoscoy, A. V., Lapizco-Encinas, B. H., *Lab on a Chip* 2010, *10*, 3235-3242.

[6] Shafiee, H., Caldwell, J. L., Sano, M. B., Davalos, R. V., *Biomedical Microdevices* 2009, *11*, 997-1006.

[7] Hagedorn, R., Fuhr, G., Müller, T., Gimsa, J., *Electrophoresis* 1992, *13*, 49-54.

[8] Asbury, C. L., Diercks, A. H., van den Engh, G., *Electrophoresis* 2002, *23*, 2658-2666.

[9] Bakewell, D. J., Morgan, H., *Ieee Transactions on Nanobioscience* 2006, *5*, 1-8.

[10] Chou, C. F., Tegenfeldt, J. O., Bakajin, O., Chan, S. S., *et al.*, *Biophysical Journal* 2002, *83*, 2170-2179.

[11] Park, K., Suk, H.-J., Akin, D., Bashir, R., *Lab on a Chip* 2009, *9*, 2224-2229.

[12] Moncada-Hernandez, H., Baylon-Cardiel, J. L., Pérez-González, V. H., Lapizco-Encinas, B. H., *Electrophoresis* 2011, *32*, 2502-2511.

[13] Martinez-Duarte, R., Gorkin, R. A., Abi-Samra, K., Madou, M. J., *Lab on a Chip* 2010, *10*, 1030-1043.

[14] Sanchis, A., Brown, A. P., Sancho, M., Martinez, G., *et al.*, *Bioelectromagnetics* 2007, *28*, 393-401.

[15] Ermolina, I., Morgan, H., *Journal of Colloid and Interface Science* 2005, *285*, 419–428.

[16] Pérez-González, V. H., Moncada-Hernández, H., Martínez-Duarte, R., Martínez-Chapa, S. O., *et al.*, *2011 AIChE Annual Meeting Conference Proceedings on CD-ROM*, AIChE, Minneapolis, Minnesota 2011.

[17] Hawkins, B. G., Huang, C., Arasanipalai, S., Kirby, B. J., *Analytical Chemistry* 2011, *83*, 3507-3515.

[18] Singh, A., Jayaram, J., Madou, M., Akbar, S., *Journal of The Electrochemical Society* 2002, *149*, E78-E83.

[19] Pohl, H. A., *Journal of Applied Physics* 1951, *22*, 869-871.

[20] Jones, T. B., *Electromechanics of Particles*, Cambridge University Press, USA 1995.

[21] Morgan, H., Green, N. G., *AC electrokinetics : colloids and nanoparticles*, Research Studies Press, Baldock, Hertfordshire, England 2003.

(8f) Temperature Measurement in a Microfluidic Device for Insulator-Based Dielectrophoretic Applications

Asuka Nakano, Kathleen Bush and Alexandra Ros, Department of Chemistry and Biochemistry, Arizona State University, Tempe, AZ

Temperature Measurement in a Microfluidic Device for Insulator-based Dielectrophoretic Applications

Asuka Nakano, Kathleen Bush, Alexandra Ros

Department of Chemistry and Biochemistry, Arizona State University, Tempe AZ, 85287

Direct current (DC) insulator-based dielectrophoresis (iDEP) has been used with cells and biomolecules such as DNA and proteins for separation, pre-concentration, and fractionation. Unlike the other existing analytical techniques, DEP response is governed by a particle's polarizability in an inhomogeneous electric field. This additional parameter space facilitates improved separation in a gel-free environment, which is of particular importance for more complex samples such as disease markers found in body fluids. DC iDEP has potential to be used as an alternative to AC iDEP since DC iDEP does not require electrokinetic and/or pressure pumps necessary in AC iDEP experiments.

The application of large DC voltage in iDEP results in heat generation known as Joule heating within the microfluidic device. This phenomenon is of great interest due to its influence on protein migration[1] as well as protein stability. In this work we present a means to measure fluid temperature in microfluidic systems by implementing fluorescent microscopy that enables dual color detection with an optical splitter and a CCD camera. Our previous work demonstrated DEP streaming of monomeric immunoglobulin G (IgG) due to positive DEP, however, unlike DNA and cells, the mechanisms of protein polarization remain less understood. Furthermore, additional electroosmotic and electrophoretic forces interplay with DEP resulting in complex protein migration behavior. Fluid temperature due to Joule heating is another factor, which has not been thoroughly investigated experimentally in iDEP. Therefore our study provides novel information to develop iDEP devices for separation, concentration, and fractionation.

In this work, we demonstrate a way to quantify the fluorescence emission ratio of two dyes[2]: Rhodamine B (RhB), a temperature sensitive dye, and Rhodamine 110 (Rh110), a temperature insensitive dye, using the same device developed for our IgG DEP experiments. While the emission intensity of RhB is proportional to the local temperature, it is biased by variations in the illuminating fluorescence light intensity. To eliminate the potential bias, we employed the two dye system and performed quantitative analysis by referencing the RhB fluorescence intensity with the Rh110 fluorescence intensity. Experiments were performed under the same buffer conditions and applied potentials to mimic the conditions of successful iDEP protein streaming. Over a period of 30 minutes, we tracked the change in temperature as well as temperature variations in different locations around the post regions and within the reservoirs.

Our preliminary experiments show that there is no significant temperature rise either within the channel or the reservoir despite minor challenges associated with these dyes such as photobleaching and non-specific adsorption of RhB to the PDMS surface. Our results demonstrate that the effects of Joule heating are small, which can be explained by two reasons. First, the temperature does not increase significantly within the channel since the bulk liquid is consistently refreshed by electrokinetic flow thus minimizing an increase in temperature. Second, the temperature within the reservoir does not vary to a large extent owing to the large solution volumes (~70 μL). Even though Joule heating is present, heat is dissipated causing the solution to remain at room temperature.

In addition to DC iDEP, our scope can be further expanded to temperature measurement within a channel with smaller structures and constrictions (i.e. nano-posts). With combination of AC iDEP, the nanostructures have potential to create larger DEP force and to immobilize proteins via DEP (DEP trapping). Thus, it is of importance to measure temperature fluctuations since a larger Joule heating effect is expected with such small constrictions. Our study therefore provides valuable information of the micro- and nano-environment, in which protein iDEP experiments are performed leading to more profound understanding of protein iDEP.

References

(1) Chaurey, V.; Polanco, C.; Chou, C.-F.; Swami, N. S. *Biomicrofluidics* **2012**, *6*, 012806–14.

(2) Ross, D.; Locascio, L. E. *Anal. Chem.* **2002**, *74*, 2556–2564.

(8g) Electrophoresis in Complex (non-Newtonian) Fluids: Theory and Experiments

Denise E. Posluszny, Lynn Walker and Aditya S. Khair, Chemical Engineering, Carnegie Mellon University, Pittsburgh, PA

The electrophoretic mobility of a colloidal particle in a Newtonian fluid is independent of both the particle size and shape in the thin Debye layer regime. Notably, electric fields are used to drive particles in "complex" (non-Newtonian) fluids whose rheology differs from the Newtonian ideal, e.g. in microfluidic technologies and capillary electrophoresis. However, there has been surprisingly little work on electrophoresis in non-Newtonian fluids. To this end, here we discuss our modeling and experimental results on the electrophoresis of charged colloidal particles immersed in complex fluids. Utilizing the Lorentz reciprocal theorem, we have developed an analytical framework to compute the electrophoretic mobility of a freely suspended charged colloid of arbitrary shape in a complex fluid possessing a shear-dependent viscosity. The mobility includes contributions from the fluid inside and outside the thin electrical double layer surrounding the particle. Notably, the latter endows the mobility with an explicit dependence on particle size, in stark contrast to electrophoresis in a Newtonian fluid. Sample calculations are performed for particles in power-law and Carreau fluids. To complement the theory, we have undertaken experiments with colloids of various sizes and surface chemistry in dilute polymer solutions. Our experiments highlight the need to carefully account for additional physiochemical effects (e.g. suppression of particle zeta potential via polymer adsorption) due to the microstructural constituents of the complex fluid. This in turn provides motivation to refinement of our theory to account for such factors.

(8h) Dielectrophoretic Isolation of DNA and Nanoparticles From Whole Blood

Michael J. Heller, Nanoengineering, University of California San Diego, La Jolla, CA, Avery Sonnenberg, University of California San Diego and Raj Krishnan, Biological Dynamics

AIChE 2012

Topical 3: 2012Annual Meeting of the American Electrophoresis Society

"T3006 Advances in Electrokinetic and Electrophoresis: Fundamentals"

Dielectrophoretic Isolation of DNA and Nanoparticles Directly from Whole Blood

Michael J. Heller1,2, Avery Sonneberg2, and Raj Krishnan3

University of California San Diego, Departments of Nanoengineering1 and Bioengineering2, Atkinson Hall Rm 2312, La Jolla, CA, 92093-0448, and Biological Dynamics3, 9393 Towne Centre Dr., San Diego, CA 92121

Abstract - Dielectrophoretic (DEP) devices and methods have been developed which allow the rapid isolation, concentration and detection of high molecular weight (hmw) DNA and nanoparticles directly from whole blood. Using special microarray devices operated at 20 volts peak-to-peak and 10 kHz, a wide range of fluorescent hmw-DNA and fluorescent nanoparticles can be concentrated into DEP high-field regions while blood cells are concentrated into the DEP low-field regions. A simple fluidic wash then removes the blood cells while the DNA and nanoparticles remain concentrated in the DEP high-field regions where they can be detected by fluorescence. Hmw-DNA could be detected at 260 ng/ml, which is a detection level suitable for cell free circulating (cfc) DNA biomarker analysis. Fluorescent 40 nm nanoparticles could be detected at 9.5 x 109 particles/ml, which is a level suitable for monitoring drug delivery nanoparticles. These DEP devices are now also being used to isolate cfc-DNA from Chronic Lymphocytic Leukemia (CLL) and other cancer patient whole blood samples. Blood sample to PCR is achieved in less than twenty minutes. The ability to rapidly isolated and detect disease related DNA biomarkers and nanoparticles from undiluted whole blood will benefit many diagnostic applications by significantly reducing sample preparation time, complexity and cost

Introduction - AC electrokinetic methods like dielectrophoresis (DEP) are well-known techniques for achieving effective separations of cells, nanoparticles, and biomolecules. Until recently, DEP technology had remained impractical for general use with complex biological samples (undiluted blood) and in high conductance solutions (>0.1 S/m). For example, earlier DEP work on separating bacteria from blood (~7-9 mS/cm) required a 50-fold dilution of the blood sample before the DEP separation process could be carried out. Many other DEP separations, including cells, virus, polystyrene nanoparticles, DNA and proteins also required low-conductance conditions. More recently, some progress has been made by several groups on carrying out DEP under high conductance conditions. We for one, have developed new DEP devices and methods that allow both hmw-DNA and nanoparticles to be isolated and detected in high ionic strength buffers (Krishnan R., Sullivan B.D., Mifflin R.L., Esener S.C., & Heller M.J., AC electrokinetic separation & detection of DNA nanoparticles in high-conductance solutions, *Electrophoresis* **29**, 1765-1774, 2008; Krishnan R. & Heller M.J., AC electrokinetic method for enhanced detection of DNA nanoparticles, *J. Biophotonics* **2**, 253-261, 2009; Krishnan R., Dehlinger D.A., Gemmen G.J., Mifflin R.L., Esener S.C., & Heller M.J., Interaction of nanoparticles at the DEP microelectrode interface, *Electrochem. Comm.* **11**, 1661-1666, 2009). Using these special microarrays with robust platinum microelectrodes over-coated with a thin hydrogel layer, DEP separations in high conductance buffers, and now whole blood, can be carried out where hmw-DNA and nanoparticles are concentrated into the DEP high-field regions over the microelectrodes, while micron size particles and blood cells are concentrated into the DEP low-field regions between the microelectrodes. Other groups have used lateral flow driven continuous DEP microseparators to separate blood cells in high conductivity media, and hybrid electrokinetic techniques (DEP, electrophoresis, electrothermal flow) to separate bacteria from blood cells under high conductance conditions.

Nevertheless, our results appear to be the first that show the isolation and concentration of hmw-DNA and nanoparticles into well defined high-field regions on the microarrays is due mainly to DEP.

Results: Figure 1 (below) shows the original electric field microarray device used for the initial DEP separation work. The expanded view shows how the DEP high-field (red) regions and DEP low-field regions (green) are created when and AC electric field is applied to nine microelectrodes. Original DC/AC electric field microarray devices were obtained from Nanogen and have one hundred 80um platinum microelectrodes overcoated with a thin hydrogel layer. Newer dielectrophoretic (DEP) microarray devices specifically designed for the isolation of cfc-DNA from high conductance buffer and blood samples were obtained from Biological Dynamics. In general, DEP experiments were carried out by adding about 20 ul of blood sample to the DEP microarray device. The DEP field was applied at 10 kHz and 20 Vp-p for 15 minutes to the nine microelectrodes. The microarray was then washed three times with 0.5x PBS with the DEP field on, and finally examined by epifluorescent microscopy.

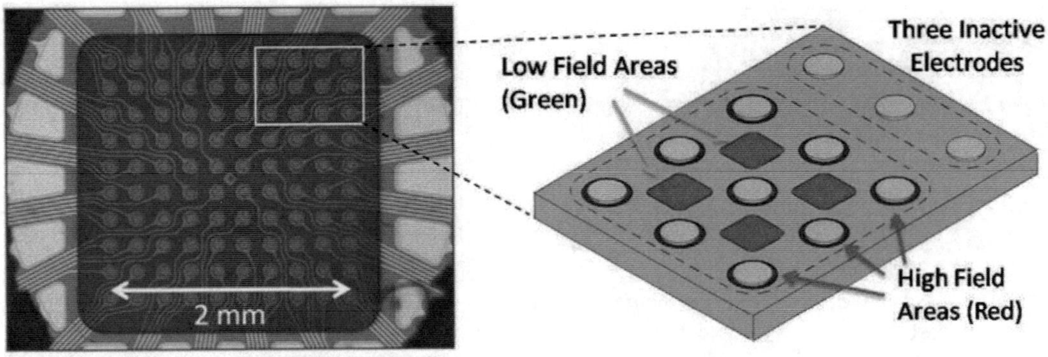

Figure 1 - DEP electric field microarray device and the DEP high-field and low-field regions

Figure 2 (below) shows the results for the DEP separation of single-stranded (ss) and double-stranded (ds) high molecular weight (hmw) DNA from whole blood and a whole blood (no DNA) control (6 to 8 mS/cm). All images are in green fluorescence. The DEP AC voltage was applied to the nine electrodes in columns 2, 3, 4, while no voltage was applied to the three electrodes in column 1, which serve as negative controls. (a) Shows the results for a 20 µl sample of whole blood containing 260 ng/ml of OliGreen stained ss-hmw-DNA, where DEP was carried out at 10 kHz and 20 Vp-p for 14 minutes, and then washed three times with 0.5x PBS. The green fluorescent rings are the hmw-ss-DNA concentrated around the edges of the activated microelectrodes, note that no fluorescence appears around the three control microelectrodes which were not activated. (b) Shows the 3D fluorescence intensity image (produced using MATLAB). (c) Shows the results for a 20 µl sample of whole blood containing 200 ng/µl of SYBR Green I Dye stained ds-hmw-DNA, where DEP was carried out at 10 kHz and 20 Vp-p for 14 minutes and then washed three times with 0.5x PBS. The green fluorescent rings are the hmw-ds-DNA concentrated around the edges of the activated microelectrodes, again note that no fluorescence appears around the three control microelectrodes. (d) Shows the 3D fluorescence intensity image. (e) Show the results for a 20 µl sample of whole blood containing SYBR Green I Dye but without any added DNA after DEP was carried out at 10 kHz and 20 Vp-p for 15 minutes and the microarray then washed three times with 0.5x PBS. (f) Shows the 3D fluorescence intensity image.

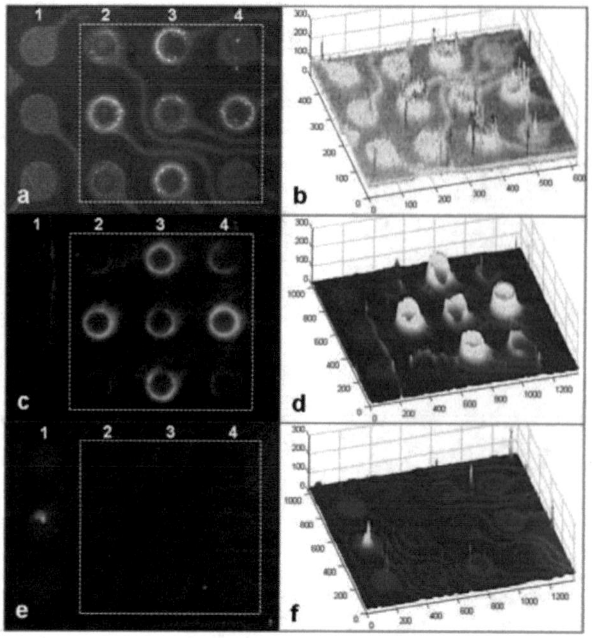

Figure 2 - Results for the DEP separation of single-stranded (ss) and double-stranded (ds) high molecular weight (hmw) DNA from whole blood, and a whole blood (no DNA) control.

Discussion and Conclusions: DEP devices and methods have been developed which have allowed the rapid isolation, concentration and detection of both hmw-ss-DNA and hmw-ds-DNA from whole blood samples. Hmw-DNA fragments could be easily detected at clinically relevant levels of 260 ng/ml for ss-hmw-DNA. In other DEP work we were able to show that fluorescent 40nm nanoparticles could be isolated and detected from whole blood at a level of 9.5 x 109 particles/ml. Additionally, using DEP devices we have been able to collect hmw-cfc-DNA from Chronic Lymphocytic Leukemia (CLL) and other cancer patient whole blood samples. Blood sample to PCR could be achieved in less than twenty minutes. Important for future diagnostic applications, these DEP results could be obtained using only very small sample volumes (20 μl whole blood) with no sample preparation being required. Further DEP experiments have demonstrated that very low molecular weight DNA appears not to concentrate into the high-field regions. This could be beneficial in that it shows, at least at the DEP AC voltages and frequencies being used, that the process is more selective for high molecular weight DNA. Thus, DEP may have a selective advantage for the isolation of cancer and other disease-related cfc-DNA, rather than the lower molecular weight apoptotic DNA normally found in blood. In addition to specific fluorescent or colorimetric stains, numerous other molecular biology techniques including PCR, *in situ* hybridization and immunochemistry detection can be easily carried out in the same sample chamber where isolation of the entities has occurred. Taken together, the results of this study support the enormous potential of DEP as a "seamless sample-to-answer" technique for the rapid isolation and detection of disease related cfc-DNA and nanoparticles directly from whole blood and other complex samples.

(8i) Pressure Generation Using Rectified AC Electroosmotic Flow with Field Effect Flow Control

Wen-I Wu and P. Ravi Selvaganapathy, Mechanical Engineering, McMaster University, Hamilton, ON, Canada

PRESSURE GENERATION USING RECTIFIED AC ELECTROOSMOTIC FLOW WITH FIELD EFFECT FLOW CONTROL

Wen-I Wu and P. Ravi Selvaganapathy

Department of Mechanical Engineering, McMaster University, CANADA

Rectified AC electroosmotic (EO) pumps using field-effect flow control (FEFC) have several advantages over DC EO pumps such as electrolysis elimination and low driving voltages[1], however it has not been fully characterized. Prior literature[1][2] focused on flow generation and calculated pressure heads based on the maximum flow rate($P_{max} = Q_{max} \cdot R$). Here, we report the first experimental demonstration of pressure generation using FEFC mechanism. EO flow is extremely sensitive to the back pressure, thus pressure conversion using the maximum flow rate could introduce certain errors and overestimate the performance of AC EO pumps.

This paper first details the development of the rectified AC EO microfluidic pump (Fig.1) with the ability to alter the cross-section area using a clamping mechanism, and then employs three methods to directly measure pressure head. The rectified flow is obtained by synchronous zeta-potential modulation with the driving potential in the microchannel. This pump is composed of 10 microchannels (L10mmW150µm) with an initial height of 20µm and is fabricated using the hydrophilic polyurethane (PU)-based microfabrication[3] to facilitate channel priming. In addition, elastic PU allows modulation of the channel cross section by external clamping which facilitates the characterization of the effect of channel cross section on the pressure generation capability of the micropump (Fig.2). Using FEFC, the flow rate and pressure output from this pump can be controlled through the driving electric field E_d (Fig.3a), applied frequency f (Fig.3b), gate potential V_g (Fig.4a), and phase lag between the driving and gate potentials Φ (Fig.4b). As seen in Fig.3, the rectified AC EO flow rate also depends on the cross-section area of the channel A which can be estimated by

$A = Q/\mu_{EO} \cdot E_d$ (μ_{EO}: EO mobility)

To characterize the pressure head, direct measurements including hydrostatic height, Boyle's law and pressure gauge were employed to eliminate the measurement errors (<10% variation). Working solutions including DI water, 0.1mM and 10mM PBS were tested. The current threshold for bubble generation was characterized so that the rectified AC EO pump is operated bubble-free. The comparison of pressure head generation is plotted in Fig.5. The maximum pressure head decreases with the applied frequency and the ionic strength of working solution as expected due to the slow EOF response time and thinner electrical double layer respectively. When frequency is low as 1Hz, the results from experimental measurement and calculation are similar, however the discrepancy increases with the applied frequency which indicates the commonly-used Q_{max} conversion could only be valid for DC or low frequency AC EOF, but not applicable for higher frequency. A maximum pressure output of 2.7kPa can be obtained from this device with 1Hz square-waveform signals, 1000V/cm driving electric field, 1500V gate potentials, and DI water as working solution without bubble generation over long periods of time (4hrs). Despite the relatively high gate potential which can be reduced substantially by thinner dielectric layer (25µm PU here), this rectified AC EO pump can be operated free-of-bubble with the capability of pressure generation, thus can be applicable for water removal in fuel cells and microfluidic control in drug delivery applications.

REFERENCES:

1. Wouden, E. *et al.* Field-effect control of electro-osmotic flow in microfluidic networks. *Colloids and Surfaces A: Physicochemical and Engineering Aspects***267**, 110-116 (2005).

2. Wu, W.-I., Selvaganapathy, P.R. & Ching, C.Y. Transport of particles and microorganisms in microfluidic channels using rectified ac electro-osmotic flow. *Biomicrofluidics***5**, 013407 (2011).

3. Wu, W.-I., Sask, K.N., Brash, J.L. & Selvaganapathy, P.R. Polyurethane-based microfluidic devices for blood contacting applications. *Lab on a chip***12**, 960-70 (2012).

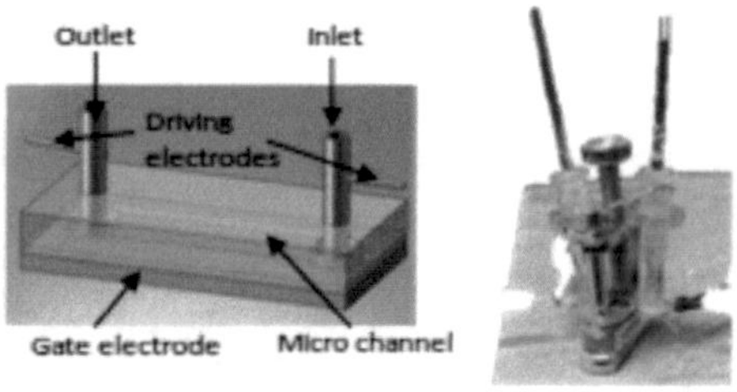

Figure 1: (a) Illustration of microfluidic pump; (b) appearance of microfluidic pump with clamp

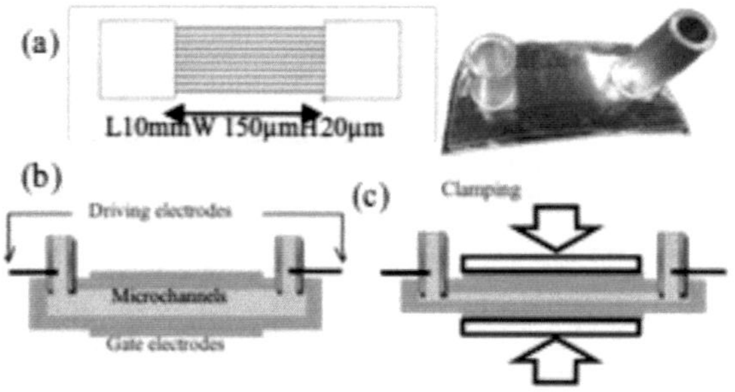

Figure 2.Schematics of (a) top view; (b) side view before clamping and (c) after clamping

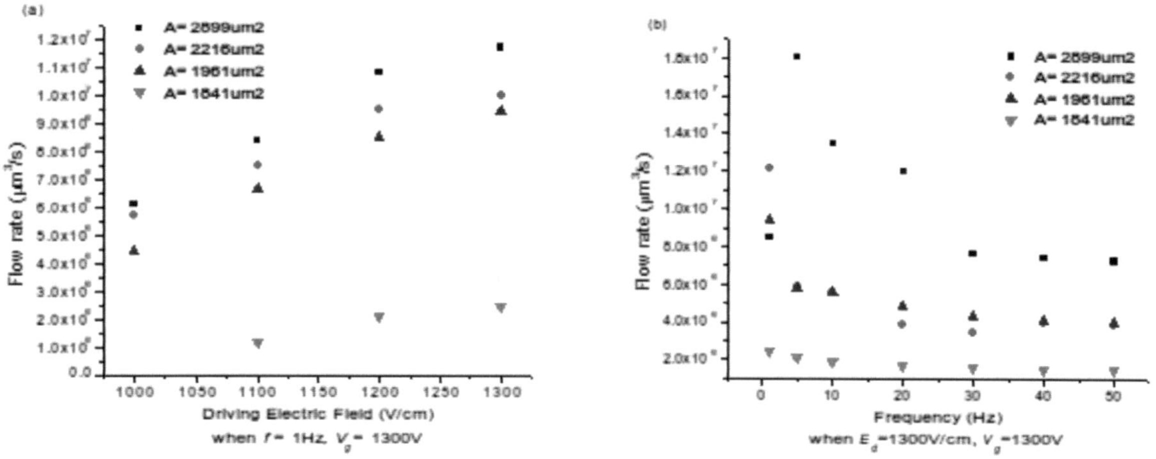

Figure 3.Effect of (a) the driving potential and (b) the applied frequency on the rectified AC EOF for various cross-section areas

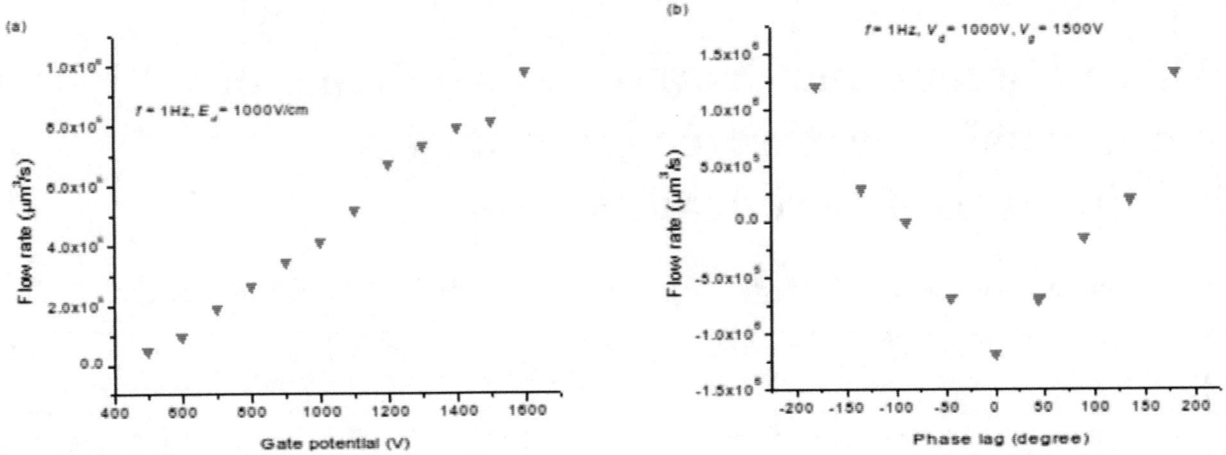

Figure 4.Effective of (a) gate potential and (b) phase lag on rectified AC EOF under the minimum cross-section area (A=1841μm²)

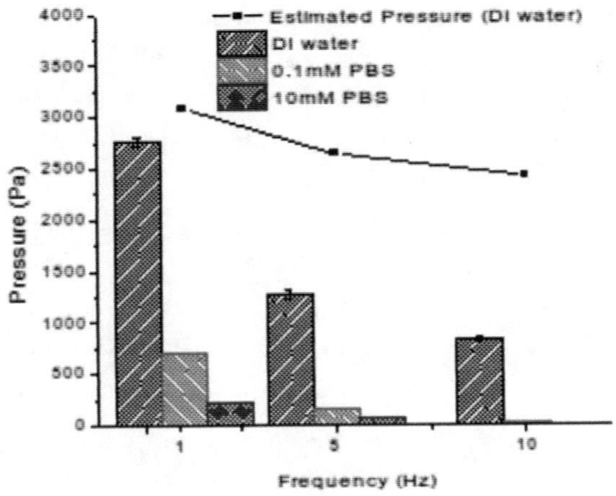

Figure 5.Pressure heads estimated by Q_{max} conversion and direct measurements at different applied frequencies when E_d= 1000V/cm and V_g= 1500V

(8j) Three-Dimensional Numerical Modeling of Electrothermal Flows in Insulator-Based Dielectrophoresis Microdevices

Akshay Kale, Saurin Patel and Xiangchun Xuan, Mechanical Engineering, Clemson University, Clemson, SC

Insulator-based dielectrophoresis (iDEP) has been established as a powerful tool for manipulating particles in microfluidic devices. In some of the most commonly encountered iDEP devices, such as the rectangular constriction microchannels, Joule Heating occurs due to the local amplification of the electric field within the constriction region. It results in an electrothermal force which manifests itself in the flow field in the form of circulations. Although the effects of Joule heating on electroosmotic flow and electrophoretic separation have been studied extensively in capillary and microchip electrophoresis, very little work has been done in analyzing Joule heating and electrothermal flow in iDEP microdevices. The previous work in this direction has focused on a two-dimensional model of the microchannel, and has not accounted for a transient behavior of the flow and temperature fields. This work demonstrates the first full-scale transient three dimensional numerical simulation of the Joule heating induced electrothermal flow in an iDEP device with a rectangular constriction microchannel in the PDMS layer. The effects on electrothermal flow will be studied for various parameters including electric field, channel dimension, solution conductivity, and PDMS thickness.

(65a) Multi-Layer Contactless Dielectrophoresis Using Thin Polyimide Films

Elizabeth Savage[1], Michael B. Sano[1], Alireza Salmanzadeh[1,2], Eva M. Schmelz[3] and Rafael V. Davalos[1], (1)School of Biomedical Engineering & Sciences, Virginia Polytechnic Institute and State University, Blacksburg, VA, (2)Engineering Science and Mechanics, Virginia Tech, Blacksburg, VA, (3)Department of Human Nutrition, Foods, and Exercise, Virginia Tech, Blacksburg, VA

Multilayer Contactless Dielectrophoresis Using Thin Polyimide Films

Dielectrophoresis (DEP), the motion of particles in a suspending medium due to their polarization in a non-uniform electric field, is an effective means of microfluidic particle manipulation. Traditionally, DEP has been performed using electrodes patterned in the bottom of a sample channel, and has been successfully utilized for a variety of applications, ranging from separating live and dead cells to trapping cells, viruses, and DNA, to determining antigen expression. However, challenges in using this technique include Joule heating, bubble formation, and electrode delamination. Contactless dielectrophoresis (cDEP) addresses these issues by using conducting fluid electrodes in side channels along a central sample channel, separated from the sample channel by a thin insulating barrier. Metal electrodes are inserted into the conducting fluid and the electric field is then applied across the sample channel without direct contact between the electrodes and the sample. When applying an electric field across the device, there is a large voltage drop across the insulating barrier and a relatively small drop across the sample channel itself. In order to achieve a sufficient voltage drop across the sample channel, much larger voltages must be applied across the device. The maximum applied voltage is limited by instrumentation capabilities and the dielectric breakdown properties of the barrier material.

The insulating barrier behaves as a resistor and capacitor in parallel and analysis shows that a reduction in resistance and an increase in capacitance of the barrier decreases the voltage dropped across this barrier and enables cDEP devices to operate over a larger frequency spectrum. For mammalian cells with similar phenotype, the largest difference in the Clausius-Mossotti factor occurs close to their first crossover frequency. In low conductivity physiologically relevant buffers, this is usually between 5-100 kHz, a range reaching below the lower frequency limit of some previous single layer PDMS cDEP devices.

A multilayer device in which the fluid electrodes and sample channel are in separate stacked layers can help achieve a reduction in the voltage drop across the insulating barrier. Increasing the area of the electrode-sample channel interface reduces the resistance while increasing the capacitance of the barrier. In single-layer devices, this interfacial area is limited by the depth of the sample channel, which in turn is limited by the depth of the silicon wafer stamps whether using deep reactive ion etching (DRIE) or photoresist spin-coating (such as SU-8). However, a multilayer device enables the fluid electrode area to be increased by locating it above the sample channel; it is then limited only by the specified channel width.

In this work, we capitalized on the observation that the voltage drop from fluid electrode to sample channel can be further decreased by choosing a barrier material that has a higher relative permittivity than PDMS. Here, a commonly available thin polyimide tape was utilized as the barrier between the fluid electrode and sample channel. The use of a higher dielectric constant self-adhesive thin polyimide film with a multilayer cDEP device provides a simple and cost-effective means of significantly increasing the potential of cDEP for sorting of mammalian cells.

Master stamps were created for the fluid electrodes and sample channel on separate silicon wafers using deep reactive ion etching. Polymer replicates were then cast in PDMS with a 10:1 ratio of polymer to curing agent. After punching access holes to each of the fluid channels, a strip of polyimide tape was applied to the PDMS replicate containing the fluid electrode channels. The fluid electrode and sample channel substrates were bonded together after exposure to air plasma for two minutes such that only the electrode cannel was in contact with adhesive. The completed devices were stored under vacuum.

Prior to experimentation, the fluid electrode channels were filled with phosphate buffered saline with a conductivity of 1.4 S/m. Polystyrene microspheres were suspended in either DI water or a low conductivity sucrose solution with conductivities of 0.001 and 0.01 S/m, respectively. A high voltage amplifier and series transformer were used to apply voltages with amplitudes up to 300 V_{RMS} between 10 and 800 kHz. Preliminary experiments match theoretical predictions that these multilayer devices with polyimide film barriers induce a DEP response in particles over a wider frequency spectrum than single layer devices with PDMS barriers.

This work presents a simplified process for fabricating multilayer cDEP devices. The use of manufactured polyimide films ensures uniform thickness of the insulating barrier between batches and provides superior electrical performance compared to PDMS. The use of an adhesive film eliminates the need for multiple bonding steps and the multilayer fabrication process allows for the use of geometries which maximize the frequency response of these devices.

(65b) Rapid, Free Solution Electrophoretic Separation of Long DNA

Angela Jones, James W. Schneider and Max A. Fahrenkopf, Chemical Engineering, Carnegie Mellon University, Pittsburgh, PA

We have developed a rapid, free solution electrophoretic DNA separation technique for long DNA (greater than 1 kb) that is capable of separating up to 6 kb with 1 kb resolution, with the potential to theoretically separate up to 48 kb with hundred-base resolution in a matter of minutes. In our system, nonionic micelles are transiently attached to the end of the DNA molecules and lower the DNA's electrophoretic mobility in a length-dependent manner. By tuning the size of the micelle, we can separate DNA in a specific size range. Because the separation takes place in free solution rather than in a gel, the separation time is also extremely low.

Traditionally, sieving matrices in the form of gels have been used in capillary electrophoresis to perform length-based separation of DNA molecules. However, long DNA molecules cannot be separated in these traditional gels because the DNA undergoes biased reptation, aligning in the direction of the electric field and migrating with a length-independent mobility. In the 1980s, pulsed-field gel electrophoresis was developed which increased the length of DNA that could be separated to the millions of base pairs, but has the disadvantage of requiring several days for separation. Techniques that have been developed in the past decade to separate long DNA include top-down approaches such as nanofabricated pillars, and bottom-up approaches such as self-assembled magnetic particles. The major drawbacks to these methods include the time consuming and expensive nature of nanofabrication processes, and the low resolution capabilities of the matrices. Instead of single base resolution achieved with traditional gels, these artificial matrices can typically only resolve DNA molecules that differ in length by thousands of bases. There are many applications such as genome mapping and DNA fingerprinting which involve separating DNA thousands of bases long, and improving the resolution of these separations is critical.

(65c) Electro-Hydrodynamic Encapsulation of Drugs Into Porous Polymer Films

Ezinwa Elele, Yueyang Shen, Ramana Susarla and Boris Khusid, Chemical, Biological & Pharmaceutical Engineering, New Jersey Institute of Technology, Newark, NJ

To eliminate adverse effects of electro-chemical reactions at the fluid-electrode interface in the course of electro-hydrodynamic (EHD) deposition, a fluid specimen does not directly come in contact with electrodes (Appl Phys Lett 97, 233501, 2010). Once deposited, a droplet spreads over and imbibed into a porous film. Compared to inkjet technologies, droplet deposition was not accompanied by fluid flow through the nozzle. Due to this feature, the EHD method provides a precise control over the deposited droplet volume and the ability to deposit highly viscous fluids. To test the method feasibility, experiments were carried out on poorly water-soluble ibuprofen and griseofulvin and porous HPMC films prepared by freeze-drying. To prepare drug solutions for printing, ibuprofen and griseofulvin were dissolved in PEG 400. Viscosities of these solutions exceeded the operating range of commercially available inkjet print heads by a factor of 15-20. The positioning accuracy of liquid droplets was about 1.2%. The relative standard deviation of the mean weight of a drug solution printed to fabricate a dosage unit was about 0.78%. The results presented show that EHD DOD offers a powerful tool for the evolving field of small-scale pharmaceutical technologies for tailoring medicines to individual patient's needs by printing a vast array of predefined amounts of therapeutics arranged in a specific pattern on a porous film.

This work was supported by the NSF ERC for Structured Organic Particulate Systems

(65d) Practical Platforms for High Throughput Sample Preparation Using 3D Carbon-Electrode Dielectrophoresis

Rodrigo Martinez-Duarte and Philippe Renaud, Microsystems Laboratory, Ecole Polytechnique Fédérale de Lausanne, Lausanne, Switzerland

The aim of this work is to develop electrokinetics-based practical platforms for a variety of applications including diagnostics, cell sorting and food safety. Therefore, the priority is on high throughput, high efficiency and user-friendliness.

The use of 3D carbon electrodes in dielectrophoresis leads to higher throughput in particle separation. 3D electrodes penetrating the bulk of the sample improve the throughput of DEP devices by minimizing the distance from a targeted particle to the closest electrode. Moreover, the use of 3D electrodes allows for the widening of the channel cross section, particularly its height, which inherently increases the processed sample volume per unit of time. Here we first present the technology development of 3D carbon-electrode DEP, from the low cost fabrication process of tall (>100 μm-high) carbon electrodes and their integration to robust microfluidics networks; to the inclusion of the DEP chips in two user-friendly portable platforms: one based on a syringe pump and another one based in centrifugal microfluidics. At last, results when using this technology in collaboration with different groups are presented and include a cell viability assay, cell selection and washing to improve sensitivity of PCR, enrichment of *M. smegmatis* towards better drugs to treat tuberculosis, DNA concentration and on-chip electrical lysis. Current work is on the separation of melanoma cells from lymphocytes.

(65e) Towards Point-of-Care HIV-1 Detection Through Electrical Sensing-On-a-Chip

Hadi Shafiee[1], Monty Jahangir[1], Fatih Inci[1], Shuqi Wang[1], Daniel R. Kuritzkes[2] and Utkan Demirci[1,3], (1)Bio-Acoustic-MEMS in Medicine (BAMM) Laboratory, Center for Biomedical Engineering, Department of Medicine, Brigham and Women's Hospital, Harvard Medical School, Boston, MA, (2)Division of Infectious Diseases, Brigham and Women's Hospital, Harvard Medical School, Boston, MA, (3)Harvard-MIT Health Sciences and Technology, Cambridge, MA

HIV diagnostics in resource-limited settings play a critical role to provide appropriate and timely care to patients. More than 95% of deaths due to infectious diseases (malaria, HIV, and tuberculosis (TB) have been reported to occur in developing countries [1-4]. 67% of the HIV-1 infected population (33.3 million people worldwide) are livening in Sub-Saharan Africa [5]. The World Health Organization (WHO) is rapidly expanding the number of AIDS patients receiving antiretroviral therapy (ART) in resource-constrained settings to control this disease. Current technical approaches for HIV diagnosis in developed world market utilizing IgG, IgG/p24, DNA, RNA, reverse-transcriptase activity biomarkers are time consuming and require sophisticated laboratory infrastructures and consequently are not compatible with resource-limited settings. These efforts, however, are significantly restricted by the prohibitive cost to implement.

ART monitoring tools, *i.e.*, CD4 cell counts by flow cytometry, and HIV viral load monitoring by reverse transcription-polymerase chain reaction (RT-PCR) are costly and require experts to implement the tests, which consequently restrict our ability to control this pandemic. Several methods have been investigated and developed to design portable CD4 cell counter tools, including electrical sensing, microfluidic lensless imaging, fluorescence staining, microscopy counting, and flow cytometry. Although CD4 cell count can be used to initiate and monitor treatment, this method cannot detect early virological failure and HIV viral load tests must be implemented. Current technologies available for viral load tests require expensive instruments and reagents, experts to conduct the tests, and advanced laboratories, which are not available at the point-of-care (POC). Thus there is an immediate and unmet need to create and develop an easy to use, portable, and inexpensive point-of-care (POC) diagnostic tool for HIV1 detection testing at resource-limited settings to monitor AIDS patients and facilitate the access to ART treatment.

In biosensor technology, microfluidics have shown promising and important advantages such as sample pretreatment including preconcentration, separation and detection, portability, handling of small sample volumes, and high-throughput analysis [6]. Due to the above advantages, one of the most promising opportunities of microfluidic platforms is reported to be in POC diagnostics and treatment monitoring applications. Biosensing technologies including optical, fluorescent, electrical and electrochemical sensing methods are compatible with microfluidic platforms. Due to easy microfabrication of interdigitated electrodes using photolithography, easy integration of microelectrodes into multi-array or microprocessor-controlled diagnostic tools, electrochemical sensing is a promising and powerful technology. This method eliminates the labeling step for sensing which consequently reduces the nonspecific detection. Furthermore, utilizing interdigitated microelectrodes, the impedance change at the surface of the microelectrodes and signal/noise ratio can be maximized and the interfering of nontarget analytes in the solution can be minimized.

Microfluidics-based electrical sensing has provided a means of probing cellular concentration, size, membrane capacitance, cytoplasmic conductivity and permittivity. Electrical measurements on mammalian cells through Coulter counting and DEP (Dielectrophoresis) have provided means of probing cellular concentration, size, membrane capacitance, cytoplasmic conductivity and permittivity. The electrical data recorded by these means can possibly be used to distinguish cells without using fluorescent or magnetic markers. Living cells are insulators at low frequencies. This monitoring has been used to study the behavior of macrophages, endothelial cells and fibroblasts. Sohn *et al.* [7] demonstrated the linearity between cellular capacitance in eukaryotic cells and the intracellular DNA content. There have been successful efforts to utilize electrical sensing using carbon nanotubes to detect single viruses [8]. However, current designs only detect a virus that passes by and are not specific to a virus and not adaptable for the POC in resource-limited settings. Also, use of magnetic particles have been attempted to attach to viral particles, however, separating nanoparticles with and without captured viruses is challenging for the counting step especially for POC applications [9].

In this study, we have successfully isolated, enriched, and detected HIV-1 particles utilizing a microfluidic device. Spiked samples with HIV1 were mixed with magnetic beads conjugated with gp120 antibodies to selectively capture target viruses. Our results show that this label-free method is capable of capturing and detecting HIV1 in a rapid, simple, and inexpensive fashion. This method can be potentially used in the detection of HIV at seroconversion (4-5 weeks) and asymptomatic (up to 12 years) period of the disease.

References:

1. Yager, P., G.J. Domingo, and J. Gerdes, *Point-of-care diagnostics for global health.* Annu Rev Biomed Eng, 2008. **10**: p. 107-44.

2. Urdea, M., et al., *Requirements for high impact diagnostics in the developing world.* Nature, 2006. **444 Suppl 1**: p. 73-9.

3. Hay Burgess, D.C., J. Wasserman, and C.A. Dahl, *Global health diagnostics.* Nature, 2006. **444 Suppl 1**: p. 1-2.

4. Girosi, F., et al., *Developing and interpreting models to improve diagnostics in developing countries.* Nature, 2006. **444 Suppl 1**: p. 3-8.

5. UNAIDS, *Report on the Global AIDS Epidemic.* 2010: p. http://www.unaids.org/globalreport/global_report.htm.

6. Wang, J., A. Ibanez, and M.P. Chatrathi, *On-chip integration of enzyme and immunoassays: simultaneous measurements of insulin and glucose.* J AM Chem Soc, 2003. **125**(28): p. 8444-5.

7. Sohn, L.L., et al., *Capacitance cytometry: measuring biological cells one by one.* Proc Natl Acad Sci U S A, 2000. **97**(20): p. 10687-90.

8. Patolsky, F., et al., *Electrical detection of single viruses.* Proc Natl Acad Sci U S A, 2004. **101**(39): p. 14017-22.

9. Chen, G.D., et al., *Concentration and purification of human immunodeficiency virus type 1 virions by microfluidic separation of superparamagnetic nanoparticles.* Anal Chem, 2010. **82**(2): p. 723-8.

(65f) Characterizing Silver Nanoparticle-Induced Modifications to the Dielectric Response of Cryptosporidia Oocysts

Yi-Hsuan Su[1], Walter Varhue[1], Kuo-Tang Liao[2] and Nathan Swami[2], (1)Electrical & Computer Engineering, University of Virginia, Charlottesville, VA, (2)Electrical & Computer Engineering Department, University of Virginia, Charlottesville, VA

Water-borne pathogens such as *Cryptosporidium parvum* and *Giardia lamblia* cause high rates of gastro-intestinal infections in the developing world, as well as in the US [1],[2]. These oocyst forming pathogens are highly resistant to traditional de-activation methodologies, such as size-exclusion filters and chlorine treatments [3]. While the infection is self-limiting in immune-competent individuals, it can be life-threatening in immune-compromised individuals, such as, people infected with the human immunodeficiency virus (HIV), where it can lead to fatalities [4]. Hence, early detection and de-activation of these water-borne pathogens at ultra-low concentration levels is a major healthcare priority. However, the lack of diagnostic technologies for rapid and highly-sensitive screening of oocyst modifications upon contact with key disinfectants has largely limited the development of de-activation methodologies for these pathogens. Immuno-fluorescence methods cannot distinguish between different species of the pathogens and cannot effectively discern infectivity. Infectivity assessment methods based on measurement of oocyst infectivity within an animal model require a rather high concentration of oocysts (~10^6oocysts/mL), and involve multiple filtration and concentration steps that do not easily allow for routine and quantitative assessment of the relative effectiveness of each disinfectant [5]. Dielectric spectral methods can distinguish speciation of the oocysts based on fingerprints in the frequency response [6], which can be correlated to animal infectivity models to obtain highly sensitive information on modifications to oocyst infectivity. Herein, under a non-uniform AC field, oocysts were polarized to experience negative dielectrophoresis (pushed away from high field region) at low frequencies (10-100 kHz) where the capacitor due to the oocyst wall determines the dielectric spectra, while at higher frequencies (>200 kHz) where the capacitor becomes conducting, the oocyst cytoplasm determines the dielectric spectra. Modification of the oocyst by disinfectants was found to affect the oocyst cytoplasm to cause it to become less polarizable. Hence, while a functional cytoplasm exhibits a strong degree of positive dielectrophoresis, successively less functional oocysts show weaker degrees of positive dielectrophoresis. To enhance the dynamic range for quantification of positive dielectrophoresis, we utilize constriction devices [7], so that the oocysts are trapped at the constriction tips under positive dielectrophoresis. In this manner, we report on the modification of the dielectrophoretic trapping force on the oocyst after various disinfectant treatments through tracking velocity profiles of the oocyst towards the constriction tip.

[1] Mead PS, Slutsker L, Dietz V et al. (1999). "Food-related illness and death in the United States". *Emerging Infect. Dis.* **5** (5): 607–25. doi:10.3201/eid0505.990502

[2] Chen, X.M., et al., "Cryptosporidiosis". N Engl J Med, 2002. 346(22): p. 1723-31.

[3] Korich, D.G., et al., "Effects of Ozone, Chlorine Dioxide, Chlorine, and Monochloramine on Cryptosporidium-Parvum Oocyst Viability". Applied and Environmental Microbiology, 1990. 56(5): p. 1423-1428.

[4] Jellison, K.L., H.F. Hemond, and D.B. Schauer, "Sources and species of Cryptosporidium oocysts in the Wachusett Reservoir watershed". Applied and Environmental Microbiology, 2002. 68(2): p. 569-575

[5] Parr, J.B., et al., Detection and quantification of Cryptosporidium in HCT-8 cells and human fecal specimens using real-time polymerase chain reaction. Am J Trop Med Hyg, 2007. 76(5): p. 938-42.

[6]] C. Dalton, A.D. Goater, J.P.H. Burt, H.V. Smith, "Analysis of parasites by electrorotation", Journal of Applied Microbiology (2004), 96, 24–32.

[7] Swami, N.; Chou, C.-F.; Ramamurthy, V.; Chaurey, V. Enhancing DNA hybridization kinetics through constriction-based. *Lab on a Chip* **2009,** *9,* 3212–3220

(65g) Dielectrophoresis for Characterizing Electrical Properties of Mouse Ovarian Surface Epithelial (MOSE) Cells

Alireza Salmanzadeh[1,2], Elizabeth Savage[1], Michael Sano[1], Mark A. Stremler[1,2], Paul C. Roberts[3], Eva M. Schmelz[4] and Rafael V. Davalos[1,2], (1)Biomedical Engineering and Sciences, Virginia Tech, Blacksburg, VA, (2)Engineering Science and Mechanics, Virginia Tech, Blacksburg, VA, (3)Department of Biomedical Sciences and Pathobiology, Virginia Tech, Blacksburg, VA, (4)Department of Human Nutrition, Foods, and Exercise, Virginia Tech, Blacksburg, VA

Lab-on-a-Chip (LOC) devices have commonly been used to isolate and uncover the intrinsic properties of rare cells. Reactions in LOC devices occur quickly and require smaller quantities of samples and reagents compared to their macroscale counterparts. Recently, a number of technologies have been developed which are capable of isolating cells based in their intrinsic properties. These techniques hold great potential as a means of early cancer detection. In this study we used contactless dielectrophoresis devices to study the electrical properties of ovarian cancer cells.

Ovarian cancer is the fourth leading cause of death in women in the United States among all cancers. The main reason for this high rate of mortality is the inability to properly detect these carcinomas early. The lack of adequate cell models to study different cancer stages and the lack of a reliable technique to isolate these cancer cells from peritoneal fluid have hindered the investigation of ovarian cancer.

Dielectrophoresis (DEP) is a commonly employed technique for cell manipulation, isolation, enrichment, and mixing in microdevices. Cell membrane morphology, membrane surface permeability, cytoplasm conductivity, radius, and surface proteins influence the DEP response of cells. Due to this, different cell types exposed to the same electric field exhibit unique DEP responses suggesting that cells can be selectively manipulated using DEP. Additionally, the influential nature of the cell membrane on DEP response provides a mechanism to calculate its electrical properties based on DEP response observations. In this work, we investigate the changes in electrical properties of mouse ovarian epithelial (MOSE) cells as the cells progress from relatively benign to aggressive stages.

A continuous sorting low frequency cDEP device was fabricated in PDMS. A silicon master stamp was first fabricated using typical photolithography techniques and deep reactive ion etching. The device was then cast in PDMS using a 10:1 ratio of monomers to curing agent. The polymer replicates were bonded to glass slides using air plasma after access holes were punched. Four different stages of MOSE cells were independently driven through the device using a syringe pump at a flow rate of 0.005 mL/hour. A 200 V_{RMS} signal was then applied across the device for frequencies between 5 and 100 kHz. The distribution of cells within the device was recorded and processed to determine the first crossover frequency. Measurements of the cell radius were then used to calculate the area specific membrane capacitance for each cell line.

We found that cell membrane capacitance increases as the cells progress into more malignant states. This is consistent with previous studies which show changes in the cells mechanical properties and membrane morphology. We hypothesize that the differences in membrane capacitance were due to morphological differences and changes in the cytoskeleton among stages of these transformed MOSE cells. This work is the first step to understanding the difference in electrical properties of cell associated with ovarian cancer and future work will focus on isolating these cells from the native cells found in peritoneal fluid.

(65h) A Microfluidic Device for Electrophoretic Trapping and Irreversible Electroporation of Bacterial Cells

Md. Shehadul Islam, Mechanical Engineering, McMaster University, HAMILTON, ON, Canada and P. Ravi Selvaganapathy, Mechanical Engineering, McMaster University, Hamilton, ON, Canada

Cell lysis or cellular disruption is a process by which cellular membrane integrity is disrupted by inducing small pores in order to release the internal contents of the cell such as DNA, RNA, and proteins. Various techniques that can be categorized as mechanical, chemical, thermal and electrical have been used for lysing cells [1]. In the electrical method, cells are lysed by exposure to an external electric field. The advantage of this method over others is that it allows lysis without the introduction of any chemical and biological reagents and also allows rapid recovery of inter-cellular organelles. When an electric field is applied, the cell membrane can act as a capacitor and a potential difference is established between intercellular and extracellular regions. This difference is known as transmembrane potential (TMP) [2] and for a spherical cell can be determined by-

TMP: $\nabla\emptyset(t) = 1.5aE\cos\partial(1 - exp(\frac{t}{t_m}))$

Where: a= radius of cell, E= external electric field, t_m= charging time of membrane.

When this potential is about 1V, the cell membrane becomes permeable as small pores are created on its surface. This process is called electroporation [2]. Depending on the field strength and duration of the field, these pores might be transient or permanent. Permanent pores can be created by increasing the intensity and exposure to the electric field. This process is known as irreversible electroporation and can be used for cell lysis. Depending on the size of the cells, electric field strength of 1000V/cm to 2000 V/cm is required to lyse bacterial cells [3].

Such high electric fields can be obtained either by applying a high potential or by reducing the gap between the electrodes through microfabrication [4]. However, microfabrication of the electrodes is expensive. In this paper, we demonstrate a method to obtain localized high electric field without the use of microfabricated electrodes. We also demonstrate eletrophoretic trapping and accumulation of cells at the localized region and subsequent lysis of those cells. The device consists of a nanoporous polycarbonate membrane sandwiched between two microchannels with electrodes embedded at the reservoirs of the microchannels. Since the resistances of the nanopores are significantly larger than that of the microchannels and the thickness of membrane is very small, any potential applied at the electrodes generates a localized high electric field. For instance, this device can generate approximately 1.8 KV into the intersection of two channels using an operating voltage of 300V. We demonstrate that there is a threshold electric field below which the device operates predominantly as an accumulator of cells suspended in the microchannel and above which it starts to lyse the accumulated cells. For instance, application of 50 V (electric field 308 V/cm) to the device with a flow rate in the channel of 100uL/hr leads to rapid accumulation of cells on top of the membrane. We use GFP expressed E. coli as well as normal E. coli with propidium iodide dye to visualize the accumulation and lysis. When GFP expressed E. coli is used, the fluorescence burst is observed when the electric field is higher than a threshold value confirming lysis. Lysis or electroporation creates pores in the cells allowing transfer of propidium iodide from the outside into the cell which then attaches to the DNA and makes the cell red. Subsequent to the accumulation when normal E. coli with propidium iodide dye is used, application of 300 V (electric field 1850 V/cm) leads to the change in color to red indicating that the cells are lysed. Cell lysis can be completed within 2 to 5 min. Apart from rapid accumulation and low voltage lysis of cells another advantage of this device may be handling of intercellular contents. Since the pore size of the polycarbonate membrane used is 400nm, intercellular organelles are retained while DNA can easily pass through them and be collected downstream. This device is useful for sample pretreatment in a micro total analysis system. As it can easily lyse bacteria, lysis of mammalian cells as well as reversible electroporation for transfection of molecules into the cell are possible by using this device.

References:

1. Ying Huang, Elizabeth L. Mather, Janice L. Bell, Marc Madou, MEMS-based Sample preparation for Molecular diagnostics, Anal Bioanal Chem, 2002, 372: 49-65.

2. Neumann, E., Sowers, A.E.,C.A.E., Electroporation and Electrofusion in Cell Biology, 1999.

3. Hsiang-Yu Wang, Arun K. Bhunia, Chang Lu, A microfluidic flow-through device for high throughput electrical lysis of bacterial cells based on continuous dc voltage, BIOSENSORS & BIOELECTRONICS, 2006.

4. Hang L U, Martin A. Schmidt and Klavs F. Jensen, A microfluidc electroporation device for cell lysis, Lab on a Chip, 2004.

(65i) Aggregation of Biofilm Forming Bacteria Induced by Insulator Based Dielectrophoresis

William Braff[1], Dana L. Willner[2], Phil Hugenholtz[3], Korneel Rabaey[4] and **Cullen R. Buie**[1], (1)Department of Mechanical Engineering, Massachusetts Institute of Technology, Cambridge, MA, (2)University of Queensland, St. Lucia, Australia, (3)Australian Center for Ecogenomics, University of Queensland, St. Lucia, Australia, (4)Ghent University, Ghent, Belgium

Insulator based dielectrophoresis (iDEP) is an attractive technique to concentrate, separate, and characterize bacteria. We've recently found that several species of bacteria known to form biofilms including *Pseudomonas aeruginosa*, *Escherichia coli*, and *Streptococcus mitis* form aggregates during iDEP trapping. In each case, strains known to form biofilms assembled into tightly bound aggregates while strains with compromised ability to form biofilms did not (or did so inconsistently). Biofilm formation is associated with increased virulence and antibiotic resistance in bacterial pathogens. Therefore, the identification of biofilm forming phenotypes bears significant clinical relevance. In this presentation we will discuss experimental results on aggregation of isogenic mutants of *P. aeruginosa* PA14. We find that iDEP can be utilized to classify cells based on both polarizability and their propensity to aggregate. Aggregation, coupled with cell polarizability, can be used to discriminate bacteria with subtle differences in cell envelope characteristics. We believe that iDEP induced aggregation represents a promising technique to rapidly obtain valuable phenotypic information on pathogenic bacteria.

(65j) Microbe Removal Using Reservoir-Based Dielectrophoresis (rDEP)

Mark Johnson[1], Robert Anderson[1], Saurin Patel[1], Jeremy Tzuen-Rong Tzeng[2] and Xiangchun Xuan[1], (1)Mechanical Engineering, Clemson University, Clemson, SC, (2)Biological Sciences, Clemson University, Clemson, SC

The removal of pathogenic microbes from water is critical to human life. In this talk we present a novel electric field-driven technique for continuous removal of microbes from water in a stacked microfluidic device. This technique, termed reservoir-based dielectrophoresis (rDEP), is based on the dielectrophoretic motion induced at the reservoir-microchannel junction as a result of the inherent size mismatch between the reservoir and the microchannel. In order to increase the throughput, multiple layers of PDMS slabs are stacked and each layer also consists of multiple microchannels. We demonstrate the continuous rDEP trapping and filtration of yeast cells from water in different regions inside the inlet reservoir simultaneously. The effects of electric field (including magnitude, AC to DC field ratio, and frequency) and suspending medium (primarily the electric conductivity) on the filtration performance are studied. A numerical model is also developed to simulate the electrokinetic transport and manipulation of cells in the stacked microfluidic device.

(85a) Surfactant Mediated Charging of Polymer Particles in Apolar Solvents

Qiong Guo, Joohyung Lee and **Sven H. Behrens**, School of Chemical & Biomolecular Engineering, Georgia Institute of Technology, Atlanta, GA

Apolar solvents do not easily accommodate electric charge, but it has long been known that surfactant additives can cause substantial particle charging in these media. While the phenomenon is commonly exploited, *e.g.* in liquid toners and electrophoretic displays, the charging mechanisms remain rather unclear. Competing popular hypotheses involve (a) the asymmetric adsorption of (reverse) charged micelles that originate from equilibrium fluctuations around a zero micelle charge, (b) the presence of dissociated ionic surfactant molecules at the particle surface, (c) the dissociation or protonation of ionizable particle surface groups, possibly facilitated by residual water introduced with the surfactant, or (d) surface charging due to acid-base interactions between the surfactant and the particle surface. Here we present measurements of the interaction energy and zeta potential of polymer particles with different acid-base characteristic and different ionizable surface groups in alkane solutions of ionic and of nonionic surfactants. Our results cannot be explained by any one of the proposed mechanisms alone, but they suggest a rather complex interplay of at least two and sometimes three different charging pathways. We will sketch a strategy to systematically disentangle the contributing effects.

(85b) Electroacoustics, Conductivity & Dispersion Forces in Nonpolar Carbon Black Dispersions

Joseph J. McDermott[1], Ian D. Morrison[1], Peter Wright[2], Jacob I Emert[3] and David A. Weitz[4], (1)School of Engineering and Applied Sciences, Harvard University, Cambridge, MA, (2)Infineum UK Ltd, Milton Hill, Abingdon, United Kingdom, (3)Infineum USA LP, Linden, NJ, (4)Department of Physics and School of Engineering and Applied Science, Harvard University, Cambridge, MA

High-volume-loading dispersions of carbon black in hydrocarbon oils are of interest both commercially, where they serve as coatings, pigment-based inks, and model systems for automotive soot; and scientifically, as they form heterogeneous, space-filling, weakly attractive gels with unique rheological properties. Common dispersants used to stabilize carbon in oils typically utilize both electrostatic and steric repulsion mechanisms. In this work, we study systems of Cabot Vulcan XC72 carbon black (a highly fractal and conductive black), stabilized using both ionic and nonionic polymer dispersants, to better understand and characterize the interparticle forces at play in these dispersions.

Characterization of the electrostatic stabilization in the dispersions involves measuring the electrophoretic mobility of the carbon particles. Because the properties of these systems are volume fraction dependent, we must make mobility measurements at the particle concentrations we are interested in studying. Instead of light-scattering-based techniques, we instead use electroacoustics to measure particle mobility. By measuring the electroacoustic response of these dispersions, along with the dispersion rheology, we can correlate the breakup of the carbon black gels with the evolution of surface charging. We additionally study the conductivity response of the dispersions to low-frequency electric fields. Because the carbon black particles are conductive, they undergo dielectrophoretic chaining in LF fields. Particle chains that are sufficiently stabilized, especially with a thick steric surface layer, tend to have lower conductivities in these fields. Finally, the sensitivity of the carbon dispersions to solution ionic concentration and valence is observed, recalling some of the earliest analytical techniques in colloid science, including examination of the Schultz-Hardy Rule for nonpolar solvents.

(85c) Ions, Ion Pairs and Inverse Micelles in Non-Polar Media

Andrei Dukhin, Dispersion Technology Inc, Bedford Hills, NY

This is review of the papers dedicated to the electrochemistry of non-polar liquids during last century. There is a list of dozens liquids that have been studied and variety of electrolytes. There is an overview of 13 different experimental methods, which have been employed for this task. Theoretical part emphasizes work done by Onsager, Debye, Fuoss, Kraus, Bjerrum and others in 1920th-30th. They had initiated and justified fundamental ideas that serve as a scientific basis for the modern handbooks on non-aqueous electrochemistry. They pointed out that electrostatic interactions lead to formation of ion-pairs and other structures in non-polar liquids. These notions and related equilibrium constrains allowed then to achieve explanation of many experiments, including conductivity dependence on concentration, dielectric constant of electrolyte solutions, deviation from the Ohm's law in non-polar liquids. They and their followers were able to extract information on ions dimension for many non-polar electrolyte solutions using conductivity and dielectric permittivity data. It was less than 1 nm in all cases. Many of these papers from 1930th and later are reviewed here.

These ideas can be applied for non-polar solutions that contain surfactants. This would logically lead to the formation of inverse micelles ions as a result of ions electrostatic interaction with dipole moments of the surfactant molecule polar part. Overview of the history associated with inverse micelles ions reveals existence of two different models for inverse micelles ions. "Symmetrical model" assumes identical size for cation and anion. "Asymmetrical model" presents cation and anion of very different sizes, one of them possibly bare. There are several arguments offered, theoretical and experimental, for supporting asymmetrical model. This model suggests existence of two different classes of surfactants depending on their role in solvating ions in non-polar liquids. Asymmetrical model explains charging of interfaces by adsorption of the smaller ion, which is supported with experiment as well.

Independently on model, inverse micelles ions must form ion-pairs. It is imperative to take into account "ions"-"ion-pairs" equilibrium in all electrodynamic effects in non-polar liquids, as was shown by Onsager. This might yield significant consequences for interpretation transient current experiments.

(85d) Effect of Thermal Treatment and Moisture On Silica Particle Charge in Non-Polar Solvents

Kwadwo E. Tettey and Daeyeon Lee, Chemical and Biomolecular Engineering, University of Pennsylvania, Philadelphia, PA

Hydrophilic colloids such as silica do not readily acquire charge in non-polar solvents and have poor colloidal stability. Surfactants, such as Aerosol-OT (AOT), are known to impart charge on these hydrophilic particles in non-polar solvents, making them stable. However, the effect of surface chemistry and water content in such non-polar media is not well understood. In this study, we examine the effect of surface chemistry and moisture on the charge of silica particles in AOT-doped toluene. The surface chemistry of silica, specifically the density of silanol groups, is varied by thermal treatment of silica particles up to 1000°C. In addition, the moisture content of the particle suspension is controlled by equilibrating them under different relative humidities. Changes in the surface chemistry as well as in moisture content significantly influence the charge of silica particles in AOT/toluene. We use Fourier transform infrared spectroscopy (FTIR), thermogravimetric analysis (TGA) and quartz crystal microblance (QCM) to gain insight into the effect of thermal treatment and moisture on the charging of silica particles. Furthermore, we relate these findings to the growth and composition of thin films formed via layer-by-layer (LbL) assembly of oppositely charged nanoparticles in non-polar solvents.

(85e) Source of Charges in Petroleum Systems

Sara M. Hashmi, Chemical & Environmental Engineering, Yale University, New Haven, CT and Abbas Firoozabadi, Department of Chemical and Environmental Engineering, Yale University, New Haven, CT

Asphaltenes, the most aromatic and heaviest weight fraction of petroleum fluids, exhibit electrostatic charging in non-polar systems. This has been evidenced in literature studies through both electrodeposition and conductivity measurements. As is the case with electrostatics in non-polar media in general, the source of charge in asphaltene systems remains somewhat mysterious.

Being highly aromatic, asphaltenes are soluble in toluene, xylene, and other aromatic solvents. In non-polar short-chain alkanes like heptane, asphaltenes precipitate from the molecular scale to full separation. A colloidal suspension phase forms during this process, allowing for measurements of the electrophoretic mobility of colloidal asphaltenes. Electrophoretic mobility measurements confirm the presence of charge on colloidal asphaltenes, while at the same time indicating the overall charge-neutrality of colloidal asphaltene suspensions. The presence of both positive and negative charges in the colloidal phase partially explains the instability of asphaltenes in heptane. At the same time, the colloidal surface charge can be tuned and manipulated through the addition of surfactants. Both ionic and non-ionic surfactants enable similar increases in average electrophoretic mobility. Ionic surfactants increase the overall charge at concentrations beyond their critical micelle concentraion (cmc) in heptane, and non-ionic surfactants increase the overall charge at concentrations less than their cmc.

In this presentation, we discuss measurements of both electrophoretic mobility abd bulk conductivity in non-polar colloidal asphaltene suspensions. We measure suspensions made by mixing heptane with isolated asphaltenes dissolved in toluene as well as by combining petroleum fluids with heptane directly. We find that the source of charge in all cases arises from the asphaltenes themselves: charge does not arise from the surfactants, regardless of whether they are present below or above their cmc in heptane. Rather, both ionic and non-ionic surfactants can enable net charging of the colloids via adsorption. We discuss the importance of both pi-pi bonding and acid-base interactions in allowing this process to occur. Furthermore we discuss possibilities for the molecular source of charges in asphaltenes, including heteroatomic and metal content.

(85f) Galvanostatic Measurements of Double Layer Formation in Doped Nonpolar Liquids

Benjamin Yezer, Department of Chemical Engineering, Carnegie Mellon University, Pittsburgh, PA, Dennis C. Prieve, Dept. of Chemical Engineering, Carnegie Mellon University, Pittsburgh, PA and Paul Sides, Chemical Engineering, Carnegie Mellon University, Pittsburgh, PA

Galvanostatic Measurements of Double Layer Formation in Doped Nonpolar Liquids

Ben Yezer, Paul J Sides, and Dennis C Prieve

Department of Chemical Engineering

Carnegie Mellon University

Pittsburgh, PA 15213

Amphiphilic surfactants are often added to nonpolar liquids to electrostatically stabilize particle suspensions in nonpolar media. The surfactant molecules form reverse micelles a few of which can acquire charges by exchanging a proton during Brownian collisions. These charged micelles then serve as charge carriers in nonpolar media and facilitate surface charges on solid particles. Previous work[1,2,3] has studied the formation of the diffuse counterion cloud when current is passed through doped nonpolar fluids held between parallel plate electrodes. In past experiments, the transient current response to a constant applied voltage is measured. This work introduces a galvanostatic method in which a potential is measured in response to a constant current through 10 μm thick solutions of OLOA 11000 in dodecane. One advantage of constant current is the easy determination of the dielectric constant of the fluid. Other fluid properties including the conductivity and total number of charge carriers are calculated from the transient potential response to a constant current. A corresponding model describes the migration of charge carriers in the cell as diffuse clouds of charged micelles accumulate at either electrode, leading to a better understanding of the dynamics of charge carriers in nonpolar fluids.

[1]D. C. Prieve, J. D. Hoggard, R. Fu, P. J. Sides, and R. Bethesa, *Langmuir* **24**, 1120 (2008)

[2]P. Kornilovitch, Y. Jeon, *J. Appl. Phys.* **109**, 064509 (2011)

[3]F. Strubbe, A. R. M. Verschueren, L. J. M. Schlangen, F. Beunis, K. Neyts, *J. Colloid Interface Sci.* **300**, 396 (2006)

(161a) Nonlinear Electrokinetics in Porous Media

Martin Z. Bazant, Chemical engineering and Mathematics, MIT, Cambridge, MA

The classical theory of electrokinetic phenomena in porous media assumes linear response to a small voltage, where the electrolyte concentration is only weakly perturbed. This talk will describe three new nonlinear electrokinetic phenomena arising in water deionization: (i) Current induced discharge of porous membranes, which leads to over-limiting current carried by co-ions and water ions (ii) time-dependent ion selectivity of capacitive charging in porous electrodes, which can be exploited for batch separations of multicomponent electrolytes, and (iii) shock electrodialysis in porous filters, due to surface conduction and electro-osmotic flow, which can be harnessed for continuous separation of salt and co-ionic particles to produce fresh water. Both theory and experiments will be discussed, as well as engineering applications to industrial wastewater treatment, such as produced water from hydraulic fracturing of oil shales ("fracking").

[Joint work with D. S. Deng, Sven Schlumpberger, Vicki Dydek, Ali Mani, Boris Zaltzman, Isaak Rubinstein, Mathias Andersen, Henrik Bruus, Michiel van Soestbergen, and P. Maarten Biesheuvel.]

(161b) The DC Force Exerted On a Charged Microparticle by an AC Electric Field

Christopher L. Wirth[1], Edmund M. Tang[2], Paul J. Sides[3] and **Dennis C. Prieve**[2], (1)Department of Chemical Engineering, Carnegie Mellon University, Pittsburgh, PA, (2)Dept. of Chemical Engineering, Carnegie Mellon University, Pittsburgh, PA, (3)Chemical Engineering, Carnegie Mellon University, Pittsburgh, PA

In 0.15 mM solution of KOH or $NaHCO_3$ in water, a single negatively charged 6-micron polystyrene sphere is levitated about 200 nm above a negatively charged planar ITO electrode by double-layer repulsion. The potential energy profile of forces acting on the sphere is determined by monitoring the distribution of elevations sampled by Brownian motion of the sphere and measured using total internal-reflection microscopy, which can detect changes in elevation as small as 1 nm. Application of a 10 kV/m electric field oscillating at 10 kHz produced oscillations in elevation which were completely swamped by Brownian motion. Nonetheless an unexpected steady attractive force was detected which was comparable in magnitude to the net weight of the sphere (0.05 pN). This additional force was proportional to the square of the electric field amplitude and is about a factor of 2 stronger in KOH compared to $NaHCO_3$. The DC force appears to be dielectrophoretic attraction [Dietz, J. Appl. Phys. 48, 1036 (1977)], resulting between aligned dipoles induced in the sphere and the planar electrode by the electric field. A similar force causes "necklaces" of colloidal particles to form.

(161c) Microfluidic Force Fields for Biochemical and Cellular Analysis

Zachary R. Gagnon, Department of Chemical and Biomolecular Engineering, Johns Hopkins University, Baltimore, MD

To date, AC electric fields have been exploited to manipulate liquid (electro-osmosis) bubbles, particles, biomolecules and cells (dielectrophoresis) in microfluidic devices. Research and application in this area, however, has been limited to the interfaces formed between two immiscible metal-liquid, particle-liquid, or gas-liquid surfaces. The influence of AC electric fields across aqueous liquid-liquid interfaces remains unexplored. As the majority of microfluidic applications involve aqueous liquid flows at low Reynolds number, fluid interfaces formed between co-flowing fluid streams are a natural occurrence in microfluidic devices. Fundamentally, many electrokinetic phenomena arise from discontinuities in ionic flux and charge accumulation at electrical interfaces. These regions of net charge interact with an external electric field and a body force is exerted on the interface.

Using a microfluidic channel with embedded electrodes, two fluid streams - one with a greater electrical conductivity, the other a greater dielectric constant - were made to flow side-by-side. An AC field was applied across the flow channel and fluid was observed to displace across the interface. The displacement direction is frequency dependent, and is attributed to the Maxwell-Wagner interfacial polarization at the liquid-liquid electrical interface. At low AC frequency (< 1 MHz), below the interfacial charge relaxation time, the high conductive stream is observed to displace into the high dielectric stream. Above this frequency, the direction of liquid injection reverses, and the high dielectric stream injects into the high conductivity stream. Hence, a liquid crossover frequency is observed and is well described by Maxwell-Wagner polarization mechanics.

Utilizing this mechanism, we explore the use interfacial polarization at liquid-liquid and polymer-electrode electrical interfaces for fluidic sensing and processing. Here, we demonstrate how these small microfluidic force fields, generated at polarized electrical interfaces, can be used to produce a portable and robust microfludic platform capable of precision fluid manipulation, electrolyte ion-sensing, particle and cell separation and solute concentration control.

(161d) Electrokinetics and High Pressure Liquid Chromatgraphy

Don Arnold, Eksigent, Dublin, CA

Fundamental investigations of electrokinetic flow led to the founding of Eksigent. From this starting point, widely varying products and applications for the origen have taken Ekisgent into high pressure liquid chromatography, chip based systems, and mass spectrometry solutions. These investigations have led to an expertise in precision flow control and fluid interfaces. Applications of these innovations range from drug discover, biomarker discovery and measurement to proteomics, genomics, and metabonomics.

Protein identification and functional analysis require new analytical tools and techniques that offer high throughput and enhanced detection sensitivity for identification of low-abundance proteins. To meet this need, Eksigent developed NanoLC™ systems with extremely precise gradient control at the 100 to 200 nL/min column flow rates preferred for nanospray MS, with integral peak parking. Multiplexing the system to allow multiple, simultaneous separations increases throughput substantially when used in combination with the cHiPLC®-Nanoflex System or with high-throughput MALDI systems, like the EKSpot™ MALDI Spotting System. Eksigent's ExpressLC® system has unique innovations to provide increased levels of chromatographic resolution and speed at low sample volumes and flow rates.or conducting library purity tests, developing validated methodologies for ADMETox studies or completing formulation and stability testing in the drug discovery and development industry.

(161e) Controlling Ionic and Water Transport Through Nanopores: Ionic Diodes, Ionic Transistors and Water Valves

Zuzanna Siwy, Physics and Astronomy, University of California, Irvine, Irvine, CA

Transport through nanopores has attracted scientific interest of researchers from various fields. Nanopores play a major role in contemporary biotechnology, because many separation and sensing processes require pores with nanometer-sized openings. Transport through pores and channels exists in virtually all biological cells and is important in such things as the regulation of heart function, nerve signals, and delivery of nutrients to the cell. My scientific interests have been focused on fabricating synthetic nanopores as small as 2 nm in diameter in order to elucidate interactions of ions and molecules with the pore structure. By increasing our understanding of these interactions we are able to develop a wide range of applications in ionic electronics for nanofluidic, lab-on-the-chip, and sensing systems. We apply these nanopores as (i) devices for controlling the flow of ions and molecules in a solution e.g. ionic diodes and transistors [1,2,3], (ii) biosensors, and (iii) valves for water [4]. Ionic diodes that we prepare consist of nanopores containing a junction between a zone with positive surface charges and a zone with negative surface charges, and offer rectification degrees of several hundreds [1,3]. Comparison of ionic diodes with pn junctions will be presented as well. Water valves use hydrophobic interactions on the nanoscale and the effect of reversible evaporation and condensation of water in hydrophobic nanopores [4].

1. I. Vlassiouk, Z. Siwy, Nanofluidic diode. *Nano Letters* **7**, 552-556 (2007).
2. E. Kalman, I. Vlassiouk, Z. Siwy. Nanofluidic bipolar transistor. *Advanced Materials* **20**, 293-297 (2008).
3. Z.S. Siwy, S. Howorka. Engineered Voltage-Responsive Nanopores. *Chemical Society Reviews*. **39**, 1115-1132 (2010).
4. M.R. Powell, L. Cleary, M. Davenport, K.J. Shea, Z.S. Siwy. Electric-field-induced wetting and dewetting in single hydrophobic nanopores. *Nature Nanotechnology* **6**, 798-802 (2011).

(185a) High Dynamic Range Proteomics

Phu Van and **Jonathan Minden**, Biological Sciences, Carnegie Mellon University, Pittsburgh, PA

Cells contain thousands of different types of proteins. These proteins exist in the cell at very different concentrations, from tens of copies per cell to millions of copies per cell, which represents a hundred-thousand-fold concentration range. The goal of comparative proteomics is to discover differences in protein expression patterns between cells, tissues and organisms grown under different conditions, with different genetic backgrounds, or at different stages of development or disease. Current proteome profiling methods are unable to detect and identify the full complement of proteins in a single experiment. This limitation is a serious impediment in many comparative proteomic analyses and is largely due to the fact that no detection system has a dynamic range that is well matched to the roughly 100,000-fold concentration range of cellular proteins. There are two general approaches to comparative proteomics experiments: peptide-centric and protein centric. Peptide-centric methods rely exclusively on mass spectrometers (MSs) for peptide identification and quantification. The dynamic range of typical MSs used for unbiased comparative proteomics is ~1,000. In protein-centric methods (which commonly involve fluorescently tagged proteins and difference-gel electrophoresis (DIGE)), protein quantification and identification are done separately by fluorescence imagers and MSs, respectively. Fluorescence imagers have a dynamic range of ~20,000. To quantify protein abundance over a 100,000-fold range, we developed a gel imaging system with a million-fold dynamic range, which is essential for detecting both low abundance proteins, such as transcription factors, and high abundance proteins, such as structural proteins, in the same experiment. To exploit the full capabilities of this new high dynamic range imaging system, we will discuss modifications to the protein labeling protocol and challenges to MS identification of low abundance proteins.

(185b) Isoelectric Focusing without Carrier Ampholytes

Frank Jahnke and Penny Ross, Sonata Biosciences, Inc., Auburn, CA

Analytical applications of isoelectric focusing use carrier ampholytes to maintain a linear pH gradient between the anolyte and catholyte. There are applications where use of carrier ampholytes is troublesome, including subsequent sample analysis by mass spectrometry.

We report here on capillary isoelectric focusing when carrier ampholytes are removed. This is motivated partially by presentations at last year's AES meeting, where results were presented and interpreted that differ considerably from ours. We report on the nature of the pH gradient measured by the change of speciation, ultraviolet absorbance and fluorescence of molecular pH probes. These results are compared with theory derived from first principles. Finally we show how single proteins and simple protein mixtures focus in such a carrier-ampholyte-free system using imaging capillary isoelectric focusing.

This text is added to conform with the stated requirement on the minimum word count of an abstract that will be accepted by the American Institute of Chemical Engineers and the American Electrophoresis Society.

(185c) Deglycosylation of Receptor Tyrosine Kinases to Improve Resolution On 2D Gels

Nancy Kendrick[1], Matt Hoelter[1], Jon Johansen[1] and Mary Ann Gawinowicz[2], (1)Kendrick Labs Inc, Madison, WI, (2)Protein Core Facility, Columbia University, New York, NY

The human genome is known to encode 58 receptor tyrosine kinase proteins (RTK) that regulate cell division and differentiation in both normal tissue and malignant tumors. Binding of a serum ligand to these high molecular weight transmembrane proteins induces protein dimerization followed by phosphorylation of tyrosine residues on the intracellular tails. RTK tyrosine phosphorylation triggers cascades of reactions that result in cell division; it is known to drives cancer growth. We have previously shown that tyrosine phosphorylation of RTKs can be detected using carrier ampholine 2-dimensional electrophoresis (2DE) in conjunction with Western blotting. However, RTK proteins are heavily glycosylated and appear as large, diffuse spots on 2D gels because of microheterogeneity of the sugar chains. The glycosylation makes it difficult to identify RTK proteins cut from 2D gels using mass spectrometry, and also difficult to distinguish between RTKs using Western blotting. Removal of the sugar chains should collapse the 2D pattern and remedy both these problems.

Deglycosylation enzymes are now commercially available. As a test, recombinant PNGase F was purchased and used to remove N-linked sugar chains from recombinant Vascular Endothelial Growth Factor Receptor (VEGF, C-terminal fragment) from ProQinase that had been glycosylated by the host sf9 insect cells. The deglycosylated pattern was compared to the original usinge 2DE. Additional treatment with O-glycosidase to remove O-linked sugar chains was performed as well as treatment with sialidase to remove terminal sialic acids. The treated and untreated samples were run on 2D gels and the patterns compared for differences. VEGF protein spot was cut from the condensed regions and analyzed using LC/MS/MS to determine sensitivity. Once conditions were optimized, human lung tumor homogenates were deglycosylated and analyzed by 2DE with phosphotyrosine Western blotting followed by protein identification using LC/MS/MS.

(185d) An Innovative LC-MS/MS Workflow for the Characterization of Combinatorial Post-Translational Modifications

Peter A. DiMaggio Jr., Department of Chemical Engineering, Imperial College London, London, United Kingdom and Nicolas L. Young, National High Magnetic Field Laboratory, Florida State University, Tallahassee, FL

Extensive genetic screens and biochemistry experiments over the past 20 years have revealed that modifications present on DNA and histones function *in concert* as docking or exclusion sites for a diverse family of proteins and protein-protein complexes that possess the activity necessary for turning the genes on or off. Indeed, many of subunits of these chromatin-associating protein complexes contain multiple putative chromatin-binding domains, thus allowing them to simultaneously recognize a particular subset of modifications to localize the activity of the complex. This apparent necessity for multiple points of contact between chromatin modifications and effector proteins suggests a highly specific regulatory checkpoint for gene activation or repression. As a result, there has been a tremendous surge in the demand for technologies that can analyze c*ombinatorial* sets of histone post-translational modifications (PTMs). Although mass spectrometry is a natural starting point for addressing such a problem, traditional LC-MS/MS platforms (i.e., reversed-phase separation and CID fragmentation) are not efficient in their ability to resolve and fragment intact modified proteins, and existing alternatives suffer from lack of throughput and reproducibility.

In this work, we present a unified experimental and computational LC-MS/MS-based workflow for the high-throughput characterization of combinatorial protein modifications. At the center of this methodology is a customized chromatography based on weak-cation exchange HILIC that provides unprecedented resolution of PTM isoforms and is directly compatible with mass spectrometry, thus allowing for the online acquisition of electron transfer dissociation (ETD) tandem mass spectra for PTM localization. A significant challenge in this workflow is the ability to interpret the resulting LC-MS/MS data, which consists primarily of 'mixed' tandem mass spectra containing fragment ions from several co-eluting modified forms that are isobaric (i.e., same exact precursor mass). Existing methods for analyzing mixed tandem MS operate exclusively on a spectrum-by-spectrum basis to determine the co-eluting species present. Here we present an innovative approach that simultaneously interprets all of the mixed tandem MS at once by directly incorporating important chemicophysical relationships derived from the chromatography into the analysis. We demonstrate the ability of this approach to produce reliable and robust tandem MS identifications over a wide dynamic range of PTM isoforms, even in the absence of 'complete' fragmentation. Several cell models will be used to illustrate the utility of the method for interrogating biological function.

(185e) Single-Stream Free-Flow Isoelectric Focusing with pH Gradients Induced by Water Splitting in Bipolar Membranes-Integrated Microfluidic Devices

Li-Jing Cheng, Chemical and Biomolecular Engineering, University of Notre Dame, Notre Dame, IN and Hsueh-Chia Chang, Chemical & Biomolecular Engineering, University of Notre Dame, Notre Dame, IN

In this abstract we report a novel microfluidic technology to generate stable pH gradients, using integrated bipolar membrane (BPM) components, for free-flow isoelectric focusing in a single-stream format. Microfluidic free-flow isoelectric focusing (FF-IEF) is a versatile and rapid preparative separation and analysis techniques in chemistry and biology. However, stable pH gradient formation is one of the practical challenges. Several techniques have been developed to produce pH gradients, including introduction of acid and base solutions and electrolysis of water. These methods, however, require either multiple input streams for specialized sheath flow or additional setup to prevent the effect of bubble generation. We have developed a simple, single-stream microfluidic pH actuation technique that directly dopes the upstream analyte flow with proton and hydroxide ions and forms a stable pH gradient in a downstream separation channel. The excess H^+ and OH^- ions are generated by electric field-enhanced water dissociation with two sets of BPM. Under a reverse voltage of 4.5 V at which the anode connects to the anion exchange side, the high electric field at the BPM junction directly splits water into H^+ and OH^- ions. Without any electron-transfer reaction to generate gas, the process produces H^+ and OH^- ions more efficiently than electrolysis. Furthermore, the injection of H^+/OH^- ions does not alter the total analyte volume and hence allows for pH adjustment without diluting analyte concentration. The microfluidic pH actuation was designed to offer efficient ion injection while insensitive to flow rate. Separation of a mixture of fluorescent pI markers has been successfully demonstrated by using the FF-IEF device under a flow rate of 10 µl/min and a separation electric field of 180 V/cm. The simple, high-speed FF-IEF separation makes the device applicable to proteomic analysis and flexible for further integration to downstream analysis tools, such as mass spectroscopy.

(185f) Rapid Detection and Quantification of Specific Proteins by Immunodepletion and Microfluidic Separation

Glareh Azadi[1], Eric Gustafson[2], Gary Wessel[2] and Anubhav Tripathi[1], (1)School of Engineering, Brown University, Providence, RI, (2)Department of Molecular and Cellular Biology , Brown University, Providence, RI

Conventional immunoblotting techniques are labor intensive, time consuming and rely on the elution of target protein after depletion. Here we describe a new method for detection and quantification of proteins, independent of washing and elution. In this method, the target protein is first captured by immunodepletion with antibody coated microbeads. In the second step, both the supernatant after immunodepletion and the untreated protein sample are directly analyzed by microfluidic electrophoresis without further processing. Subsequently, the detection and quantification are performed comparing the electropherograms of these two samples. This method was tested using an *Escherichia coli* lysate with a FLAG tagged protein and anti-FLAG magnetic beads. An incubation of as little as one minute was sufficient for detectable depletion (66%) by microchip electrophoresis. Longer incubation (up to 60 minutes) resulted in more depletion of the target band (82%). Our results show that only 19% of the target is recovered after elution from the beads. By eliminating multiple wash and elution steps, our method is faster, less labor intensive, and highly reproducible. Even in case of non-specific binding at low concentrations, the target protein can be easily identified. This work highlights the advantages of integrating immunodepletion techniques on a microfluidic platform.

(185g) The Use of Microchannel Electrophoresis to Detect Early Stages of Amyloid-Beta Aggregation

Elizabeth Pryor, Chemical Engineering, University of Arkansas, Fayetteville, AR, Christa N. Hestekin, Ralph E. Martin Department of Chemical Engineering, University of Arkansas, Fayetteville, AR and Melissa Moss, Chemical Engineering, University of South Carolina, Columbia, SC

Amyloid proteins are involved in a variety of diseases including Alzheimer's disease, Parkinson's disease, and diabetes. A number of technical challenges exist for the quantitative analysis of the different sizes of amyloid proteins which are present during the early stages of aggregation, therefore requiring new technologies to be developed for enhanced protein separations. Microchannel electrophoretic techniques have emerged as powerful tools for the quantitative analysis of proteins. Previously, we investigated the ability of capillary electrophoresis (CE) to monitor insulin oligomerization and aggregation over time. In this study, we investigated the ability of CE to monitor the early stages of amyloid aggregation using Amyloid-Beta as a model amyloid-forming protein. The sizes of Amyloid-Beta aggregates produced using size exclusion chromatography-purified Amyloid-Beta monomer as the starting material are compared to the sizes produced using a more traditional Amyloid-Beta sample preparation. Furthermore, the utility of microchip electrophoresis to detect Beta-sheet possessing species is investigated.

(185h) Evaluation of Stain-Free Chemistry for Gel Electrophoresis Based Shotgun Proteomic Workflows

Sricharan Bandhakavi[1], Tim Wehr[1], Todd Markowski[2], LeeAnn Higgins[2], Xiaoyi Xu[1], Aran Paulus[1] and Christopher Belisle[1], (1)Bio-Rad Laboratories, Hercules, CA, (2)University of Minnesota, Minneapolis, MN

In processing their samples for shotgun proteomics, researchers commonly use SDS-PAGE/gel electrophoresis to initially resolve complex biological samples. Subsequently, the gel is stained for visualization of protein lane(s), and sliced into discrete 'fractions' along each lane. Finally, each fraction is processed for in-gel protein digestion, peptide purification, and LC-MS/MS analysis. This workflow, in its entirety, is commonly referred to as 'GeLC-MS' and widely used for identifying resident proteins within complex samples of interest.

Several reports have demonstrated that compared to 'in-solution' protein digestion and fractionation, GeLC-MS enables higher total protein identifications with particular benefits for membrane/hydrophobic proteins *(Fang et al., 2010 J. Proteome Res.*, *Piersma et al.* 2010 *J. Proteome Res.*). In spite of these inherent benefits, improvements in the traditional GeLC-MS workflow (eg: in terms of speed and ease of processing) are needed to widen its adoption among researchers. Here, we investigated the suitability of tris-glycine extended-life formulation (TGX) gels with 'stain-free chemistry' (a photochemical reaction 'in-gel' allowing for proteins to be visualized by fluorescence) for faster and simplified GeLC-MS analysis of a complex biological sample.

Identical amounts of mouse 3T3 whole cell lysate were resolved using traditional Tris-HCl (4-15%) versus TGX stain-free (4-15%) gels and processed via GeLC-MS. Compared to Tris-HCl gels, TGX stain-free gels 'ran' faster allowing electrophoresis to be completed in half the time. For subsequent visualization of protein lanes, Coomassie staining/destaining was used for Tris-HCl gels, whereas a simple 30 second UV transillumination step was sufficient for TGX stain-free gels. Fractions were collected across each gel lane and proteins identified via analysis of purified, fractionated tryptic peptides on a Velos-Orbitrap mass spectrometer.

Nearly identical numbers of proteins, peptides, and spectra were identified at identical spectral identification rates using either Tris-HCl or TGX stain-free gels. Thus, incorporating TGX stain-free gels represents a valid approach for simplifying current GeLC-MS worflows used in shotgun proteomic studies. Details of our study and a simplified GeLC-MS workflow using TGX stain-free gels will be presented.

(185i) Protein Streaming Via Insulator-Based Dielectrophoresis in a Microfluidic Platform

Asuka Nakano, Fernanda Camacho-Alanis, Tzu-Chiao Chao and Alexandra Ros, Department of Chemistry and Biochemistry, Arizona State University, Tempe, AZ

Protein Streaming via Insulator-based Dielectrophoresis in a Microfluidic Platform

Asuka Nakano, Fernanda Camacho-Alanis, Tzu-Chiao Chao, Alexandra Ros

Department of Chemistry and Biochemistry, Arizona State University, Tempe AZ, 85287

Rapid and reliable separation and analysis of proteins is of great interest. This becomes especially challenging when only small sample volumes are available, concomitantly with low concentrations of proteins. Time critical situations such as surgeries and rapidly degrading samples pose additional challenges. Due to these challenges, conventional macro-scale separation techniques reach their limitations. While microfluidic devices require only pL-nL sample they offer several advantages such as speed, efficiency, and high throughput.

Here, we propose to elucidate the capability to manipulate proteins in a rapid and reliable manner with a novel migration technique, namely dielectrophoresis (DEP). We demonstrate a detailed study of various factors influencing protein DEP, employing immunoglobulin G (IgG) under DC conditions. DEP has been extensively employed as a separation, fractionation, and pre-concentration technique for large biological objects such as cells and recently for DNA. In contrast to the other existing separation techniques, DEP relies on a particle's polarizability in an inhomogeneous electric field. This additional parameter has great potential to improve separation when employed as a primary separation method as well as a component of a multi-dimensional separation. Improved separations are in demand especially when the sample becomes complex. To generate a non-uniform electric field, we specifically utilize insulator-based DEP (iDEP) where the dielectrophoretic response of proteins is provoked in a tailored microfluidic device with integrated arrays of insulating posts. Unlike DNA and cells, protein DEP behavior is not well understood; therefore our detailed study of protein DEP provides novel information to eventually optimize this protein migration method for pre-concentration and other analytical techniques.

First, we performed numerical simulations to predict protein migration behavior considering electrokinesis, protein diffusion, and DEP. With small sized (~nm) proteins, high electric field strength as well as corresponding large electric field gradient needs to be created. For this purpose, multiple post geometries were tested by numerical simulation and as a result, we found the maximum field strength of 8×10^5 V/m and maximum E^2 value of 3×10^{17} V^2/m^3 with the triangular post geometry. Based on the calculated electric field, E, and E^2, we simulated the protein concentration profiles and found that the protein concentrates in streamlines between the rows of posts in the positive DEP case, whereas protein is depleted at the exact same regions for negative DEP.

Experimentally, we investigated the DEP behavior of IgG under various buffer conditions differing in pH and conductivity. We observed streaming DEP only at pH 6.5~8 with a maximum DEP streaming enrichment of 70 %. This concentration profile was in excellent qualitative agreement with numerical simulations performed with the assumption of monomeric IgG species. Furthermore, pH dependent DEP streaming is also in agreement with the numerical simulations suggesting that the variation of protein charge and electroosmotic flow influences protein DEP streaming. Additionally, we observe micelle induced negative DEP streaming for proteins, which is also in agreement with numerical simulations. This micelle formation was further confirmed by dynamic light scattering experiments. Our study thus provides valuable information to eventually improve novel protein DEP devices for separation, pre-concentration, and fractionation.

(185j) In Silico Discovery of Biomarker Proteins for Periodontitis Using High-Throughput Proteomics and Mixed-Integer Linear Optimization

Richard C. Baliban[1], Dimitra Sakellari[2], Zukui Li[1], Yannis Guzman[1], Peter A. DiMaggio Jr.[3], Benjamin A. Garcia[4] and Christodoulos A. Floudas[1], (1)Chemical and Biological Engineering, Princeton University, Princeton, NJ, (2)Department of Preventive Denistry, Periodontology, and Implant Bioligy, Aristotle University of Thessaloniki, Thessaloniki, Greece, (3)Department of Chemical Engineering, Imperial College London, London, United Kingdom, (4)Department of Molecular Biology, Princeton University, Princeton, NJ

The search for biomarkers which can act as predictors of periodontal disease at the initiation and progression stage of periodontitis has received considerable interest during the last decade [1]. The diagnostic potential of Gingival Crevicular Fluid (GCF) has been extensively investigated due to the possibility of non-invasive collection and the complexity of molecules that it contains [2]. GCF has been shown to be the transudate of gingival tissue interestitial fluid, but during periodontal disease it is transformed into inflammatory exudate which reflects the composition of serum and includes substances derived from the structural tissues of the periodontium and oral bacteria colonizing the gingival pocket [3]. Several substances (up to 90) including cytokines, proteolytic enzymes, bacterial-derived metabolites, or products of tissue degradation have been investigated as possible indicators or predictors of disease activity, but currently no chairside tests exist that can be reliably applied for accurate diagnosis of prognosis in clinical practice.

In this talk, we present a comprehensive proteomic analysis of GCF samples to uncover candidate biomarker proteins for the detection of periodontal status using the PILOT_PROTEIN webtool [4]. A complete list of human and bacterial proteins was compiled from the analysis of 24 GCF samples from a mixture of 12 healthy and 12 diseased patients [5]. Strong candidate human and bacterial biomarker proteins were identified for further analysis. A mixed-integer linear optimization model was then developed to identify the optimal combination of biomarkers which could clearly distinguish a blind subject sample as healthy or diseased. To adequately train the optimization model, an additional 31 GCF samples from 14 healthy and 17 diseased patients were collected and analyzed with PILOT_PROTEIN.

A thorough cross-validation of the model capability was performed by selecting a training set of size *N* from the 55 samples, optimizing the biomarker selection, and then testing on the remaining 55 - *N* samples. One hundred random combinations of training sets were used to validate the accuracy of the model on the test sets. The mathematical model was able to consistently achieve a high degree of accuracy (i.e., greater than 90%) when annotating the testing set samples as healthy of diseased. The effect of training sample size was investigated and it was found that the model was able to maintain a high accuracy when utilizing a training set containing as few as ten samples. The model was then trained on all 55 samples and tested on two different blind sets containing 20 samples and 21 samples. Using an optimal combination of 6 healthy proteins and 2 bacterial proteins, the mathematical model was able to provide a perfect separation of healthy and diseased samples from both test sets.

[1] Champagne, C. M., Buchanan, W., Reddy, M. S., Preisser, J. S., Beck, J. D. & Offenbacher, S. (2003). *Periodontology 2000* **31**, 167.

[2] Buduneli, N. & Kinane, D. F. (2011). *Journal of Clinical Periodontology* **38**, 85.

[3] Delima, A. J. & Van Dyke, T. E. (2003). *Periodontology 2000* **31**, 55.

[4] Baliban, R. C., DiMaggio, P. A., Plazas-Mayorca, M. D., Garcia, B. J. & Floudas, C. A. (2012). *Analytical Chemistry*, submitted.

[5] Baliban, R.C., Sakellari, D., Li, Z., DiMaggio, P.D., Garcia, B.J., & Floudas, C.A. (2011). *Journal of Clinical Periodontology* **39**, 203.

(224a) Hydrodynamic Instabilities of Chemotactic Bacteria

Donald L. Koch, Chemical and Biomolecular Engineering, Cornell University, Ithaca, NY and T.V. Kasyap, School of Chemical and Biomolecular Engineering, Cornell University, Ithaca, NY

Over the past decade, experimental studies of swimming micro-organisms and numerical simulations of hydrodynamically interacting self-propelled particles have shown that such living fluids exhibit collective dynamical motions on length scales large compared with the size of an individual swimmer even in the absence of any hydrodynamic or chemical gradients. A continuum theory showed that living fluids consisting of pushers such as bacteria are subject to a hydrodynamic instability that involves the coupling of the shear motion caused by the active stress produced by the swimmers and their orientation distribution. Recent experiments, however, suggest that even more vigorous convective motions may arise in bacterial suspensions subject to chemical gradients. In this talk we discuss a new mechanism of instability that arises from the coupling of the bacteria concentration and the competition between chemotactic driven and the convective motions produced by an active stress. The stability of a suspension of chemotactic bacteria confined in an infinitely long channel and subjected to a stationary, linear chemo-attractant gradient is investigated. While swimming, individual bacteria exert a force dipole on the fluid which at the continuum level leads to a stress depending upon the bacterial orientation and number density fields. The presence of the attractant gradient causes bacteria to tumble less frequently when swimming along the gradient leading to a mean orientation and a non-zero chemotactic drift velocity. A balance of the fluxes due to chemotaxis and the random run-tumble motion of bacteria yields an exponentially varying number density profile across the channel in the base state. The associated bacterial stress field is also exponentially varying and is normal. This spatially non-uniform base state is unstable to fluctuations in the bacterial concentration exceeds a critical value determined by a Peclet number defined as $Pe = U_0 H/D$ where U_0 is the chemotactic velocity, D the run-and-tumble diffusivity and H the channel thickness. The instability resulting from the coupling between the active stress driven fluid flow and the bacterial concentration manifests as rectangular convection patterns driven by spanwise bacterial clusters.

(224b) Occlusive Flow of a Red Blood Cell

Thierry Savin, ETH Zürich, Zürich, Switzerland and L. Mahadevan, Harvard University, Cambridge, MA

When red blood cells (RBCs) move through narrow capillaries in the microcirculation, they deform as they flow. In pathophysiological processes such as sickle cell disease and severe malaria, RBC motion and flow are severely restricted. This can lead to vaso-occlusive crises with dramatic consequences. To understand the dynamics close to the threshold of occlusion, we used a combination of experiment and theory to study the motion of a single swollen RBC through a narrow glass capillary of varying inner diameter. By tracking the motion of the squeezed cell as it is driven by a controlled pressure drop, we measure the RBC velocity as a function of the pressure gradient as well as the local capillary diameter, and find the velocity of the cell varies with pressure head as a power-law that depends upon the lubrication gap between the cell and the capillary wall, consistent with a simple elasto-hydrodynamic model. Our study highlights the role of lateral confinement in the nearly arrested pressure-driven flow of soft confined objects and allows us to quantify the increase in the effective viscosity associated with this regime.

(224c) The Continuous Chromatographic Separation of Molecules/Particles Using Optical Electric Fields

Nicolas Alvarez[1], Claus Jeppesen[2], Kresten Yvind[2], Iwao Teraoka[3], N. Asger Mortensen[2] and Ole Hassager[1], (1)Department of Chemical and Biochemical Engineering, Technical University of Denmark, Kgs. Lyngby, Denmark, (2)Department of Photonics Engineering, Technical University of Denmark, Kgs. Lyngby, Denmark, (3)Chemical and Biological Science, Polytechnic Institute of New York University, Brooklyn, NY

We introduce a new field-flow fractionation technique, whereby molecules are separated based on their differential interaction with optical electric fields, i.e. electric fields with frequencies in the visible and near-infrared range. When solute particles are introduced to electric field gradients they are polarized based on their size and refractive index. The resulting attractive or repulsive polarization potential influences a solute particle's lateral position with respect to the gradient in the electric field. By coupling well described optical electric fields with a non-uniform flow, we demonstrate how such potentials can be used to continuously separate nanometer scale solute particles in a flowing two dimensional microchannel. Comparisons are made between theoretical axially uniform optical fields and numerical simulations considering optical electric fields with finite width. We discuss the impact of the governing dimensionless groups on the resolution of separation and the dispersion of solute molecules after separation. An experimental apparatus is introduced and comparisons are made between theoretical and experimental results.

(224d) Micro and Nanoparticle Trapping and Manipulation with Fluid Flow

Melikhan Tanyeri and **Charles M. Schroeder**, Chemical and Biomolecular Engineering, University of Illinois at Urbana-Champaign, Urbana, IL

The ability to trap and control the motion of single particles has led to key advances in science and engineering. Particle trapping is a pervasive technology that underlies nanoscience and biological analysis. In this presentation, I will introduce two new microfluidic methods based on the sole action of hydrodynamic forces for the confinement and manipulation of micro- and nanoscale particles in free-solution. Both trapping methods utilize an automated feedback-control mechanism for active control of fluid flow within microchannels using an integrated on-chip metering valve. The first method is the "**hydrodynamic trap**" where a target particle is confined at a microchannel junction by actively controlling the stagnation point of a planar extensional flow field This method enables trapping, 2-D manipulation, stretching and sorting of objects ranging from single molecules to individual cells using the sole action of fluid flow. The second method is the "**microfluidic Wheatstone bridge**" which facilitates rapid sampling of particles, cells and droplets in continuous flow for monitoring and analysis. This dynamic sampling method allows for "on-demand" trapping of multiple particles at the "bridge" channel without stopping or diverting the main sample stream. Currently, we are pushing the limits of these methods for confining small nanoparticles (< 50 nm) and large macromolecules. We anticipate that these microfluidic-based techniques will enable new scientific studies in the fields of catalysis, colloidal science, polymer dynamics, single molecule studies, systems biology, and cellular mechanics.

1. M. Tanyeri, E. M. Johnson-Chavarria and C. M. Schroeder, Applied Physics Letters (2010) **96**, 224101.

2. M. Tanyeri, M. Ranka, N. Sittipolkul and C. M. Schroeder, Lab on a Chip (2011) **11**, 1786.

(224e) Mapping Chaotic Flow States in Microscale Rayleigh-Bénard Convection to Identify Regimes of Accelerated DNA Replication

Aashish Priye and Victor M. Ugaz, Artie McFerrin Department of Chemical Engineering, Texas A&M University, College Station, TX

We have previously shown that chaotic advection in micro-scale flow geometries can be harnessed to greatly enhance the rate of thermally activated biochemical reactions (e.g., the polymerase chain reaction (PCR)). Here we present a 3D analysis of coupled flow and reaction that provides a time-resolved mapping of the DNA replication process as it unfolds. Chemical reaction kinetics are incorporated into the 3D flow model, enabling us to obtain a measure of global DNA replication rate within the entire reactor volume, expressed in terms of a characteristic doubling time. Mass action kinetics were assumed for the denaturing, annealing, and extension steps. Annealing and extension were modeled as bimolecular reactions to capture the contribution of the primers and dNTPs. Reversibility of denaturing and annealing reactions was also taken into account. Reaction models of varying complexity were compared via addition of side reactions. individual stream traces were examined for different geometries, expressed in terms of their aspect ratio (height/diameter, *h/d*). The residence time of a fluid element in a particular reaction zone and the number of times that fluid elements pass through the zones were determined for 300 randomly chosen stream traces yielding a distribution of residence times and frequencies of ingress in each of the reaction zones. At larger values of *h/d* we find that individual fluid elements on average spend more time in each temperature zone. But the number of times a fluid element passes through each zone is much larger in wider geometries (lower *h/d*), owing to the chaotic characteristics of the flow field. Since our experiments show much faster DNA replication at low *h/d*, the number of excursions into each reaction zone appears to play a more important role than the transit time through each temperature range. Finally a reaction map was constructed to determine the cylindrical cell geometries and other parameters that should be used to enhance the efficiency of microscale Rayleigh Bénard convection PCR. These results highlight the importance of the flow field's 3D characteristics.

(224f) Electrokinetically Driven Micro-Turbulence in Microfluidics with Re in the Order of 1

Guiren Wang[1], Fang Yang[2] and Wei Zhao[1], (1)University of South Carolina, Columbia, SC, (2)Department of Mechanical Engineering & Biomedical Engineering Program, University of South Carolina, Columbia, SC

We report an electrokinetically driven micro-turbulent flow phenomenon in a microchannel, where the Reynolds number (Re) based on bulk flow and channel width is in the order of 10. As we know the critical Re in a channel flow is about 2000 and a pressure driven flow with Re in the order of 10 is definitely laminar, hence, the fluid mixing is very flow. Although there have been electrokinetical micromixers published that can increase mixing, the flows are believed not to be turbulence, but chaotic. Here we design a novel electrokinetically forced pressure driven flow with Re in the order of 10 to achieve turbulence like flow. To confirm that the flow is turbulence, Laser Induced Fluorescence Photobleaching Anemometer (LIFPA) is used to measure the corresponding flow velocity features. LIFPA enables ultrahigh temporal and spatial resolution measurement. We report the measured flow features: high dissipation and diffusion rate, multiscale continuous power spectra and 3-D flow. The results open a new view of turbulence at microscale.

(224g) Mixing Enhancement in Three-Dimensional Helical Microchannels

Shahab Shojaei-Zadeh, Rutgers, The State University of New Jersey, Piscataway, NJ and **Michael C. Fechtmann**, Mechanical and Aerospace Engineering, Rutgers, The State University of New Jersey, Piscataway, NJ

We present experimental and numerical results on mixing performance for the flow of two miscible liquids in a three-dimensional helical microchannel. A novel yet straightforward method is developed to fabricate 3-D helical microchannels with different pitch length in a polydimethylsiloxane (PDMS) mold. This geometry provides a constant Dean number flow, due to the constant radius of curvature of the helix, which differentiate it from the existing planar spiral microchannels. The mixing performance is characterized, using digital image processing techniques, based on the mixing length which is defined as the stream-wise distance in the microchannel where the mixing has reached 90% of the full intensity. Experimental results show a minimum mixing length vs. Reynolds number as a common feature for all the microchannels of similar diameter and different pitch length. Increasing the Reynolds number beyond this minimum mixing length adversely influences the mixing performance by increasing the measured mixing length due possibly to the decrease in the residence time of the two streams in the microchannel. Numerical simulations are presented to support the experimental observations.

(224h) Two Phase Microfluidics with Molten Polymers

Shu-che Peng[1], Shailesh Nagarkar[1], Justin Lowen[1], Alex Boardman[1], Steven D. Hudson[2], Kalman Migler[2], Doyoung Moon[3] and **Sachin Velankar**[4], (1)University of Pittsburgh, (2)Polymers Division, National Institute of Standards and Technology, Gaithersburg, MD, (3)National Institute of Standards and Technology, (4)University of Pittsburgh, Pittsburgh, PA

Drop-based methods for fabricating complex particles in microfluidic devices have been studied heavily over the past several years. Typically, a two-phase oil/water flow is set up in a microfluidic device under conditions where capillary breakup leads to controlled generation of drops. Subsequent solidification of the drops yields particles. These approaches suffer from material limitations: they rely on fluids that can be rapidly solidified, often by UV crosslinking. We seek to overcome these material limitations, specifically we aim to unite microfluidics with polymer processing to develop a process that can fabricate polymer particles out of virtually any thermoplastic polymer.

Handling molten polymers in microfluidic devices requires entirely new concepts in device design. A new high-temperature microfluidic platform based on micromachined metal foils has been developed. The flow of the molten polymer is driven by gas pressure rather than positive displacement pumps. This talk will show that – despite the high viscosity and low interfacial tension of molten polymers – capillary breakup is still a viable route to form drops in microfluidic devices. However the mode of drop breakup is quite different in polymeric systems as compared to oil/water systems. Furthermore, the parameter space (i.e. pressure ranges) in which drops can be generated is relatively narrow, and drop sizes are highly sensitive to the operating conditions. A simple "circuit" model is developed which successfully explains these latter observations and points to operating guidelines for using pressure-driven flow in two-phase microfluidics. In summary, the vast developments in drop-based microfluidics can be transplanted to molten polymers, thus permitting the fabriction of complex particles out of a variety of thermoplastic polymers.

(224i) Vector Separation of Suspended Particles Using an Array of Slanted Open Cavities

Jorge A. Bernate[1], Chengxun Liu[2], Liesbet Lagae[2], Konstantinos Konstantopoulos[1] and German Drazer[1],
(1)Chemical and Biomolecular Engineering, Johns Hopkins University, Baltimore, MD, (2)IMEC, Leuven, Belgium

We present a microfluidic device that takes advantage of the flow characteristics in a channel in which the bottom surface is patterned with slanted ridges that create open cavities, to passively and continuously fractionate suspended particles based on their size and settling velocity. We exploit the flow that is created along the cavities to induce the vector separation of suspended particles with dimensions comparable to the height of the ridges and that move in the vicinity of the patterned surface. The width and height of the ridges are much smaller than those of the main channel to reduce the effect of recirculation in the bulk. We shall show that two regimes can be distinguished depending on the settling velocity of the particles compared to their average velocity across the cavities imposed by the fluid flow. When these velocities are similar, heavier particles settle deeper into the open cavities and exhibit larger displacements along these cavities. On the other hand, if sedimentation is negligible, smaller particles are advected by the flow deeper into the cavities and therefore deflect more than larger particles. First, we illustrate the flow characteristics by numerically solving the particle-free flow. We then present experiments separating silica and polystyrene beads of the same and different size. We also show the potential of this method to separate cells showing that different blood components can be deflected to a different extent. Finally, we discuss the potential versatility of this platform using force fields other than gravity, e.g., electric, dielectrophoretic, and magnetic, to manipulate the settling velocity of the particles. In particular, we show the enrichment of rare blood cells by depleting magnetically labeled leukocytes.

Role of Electro-osmosis in Microchannel-Nanochannel Impedance Response.

Electrical Impedance Spectroscopy has been used extensively for the experimental study and characterization of ion-exchange membranes. The impedance characteristics and the appearance of a tangent line at an apparently universal 45 degree angle for the so-called Warburg branch of the impedance plane (Nyquist) plot are predicted and understood in terms of equivalent circuit models and have been derived from fundamental electro-diffusion . However, despite the myriad of current applications, a similarly comprehensive study of microchannel-nanochannel interfacial impedance response has not been undertaken. Here we present some fundamentally new experimental results concerning a deviation from the constant-angle Warburg response seen in membrane systems, increasing with applied current in microchannel-nanochannel devices. The experiments are analyzed in terms of a fundamental theory relating some of these differences in response to the presence of electro-osmotic flow in the microchannel-nanochannel system.

A simple 1D model based on the Poisson-Nernst-Planck-Stokes equations is employed to describe the transport phenomena in a microchannel-nanochannel device. Local electroneutrality is assumed for the PNP model and the net (cross-sectional average) fluid flow is assumed to be determined by the nanochannel EOF alone. This flow is assumed Hele-Shaw flow due to the long rectangular cross-section of the channel and given by a Darcy's law type approximation to leading order in the channel depth with an electric body force. The nanoslot is modeled using a modified model for an ideally charge selective membrane which permits fluid flow and field and flux focusing effects are included via a simple geometric factor to account for different flux densities. Using these simplifications, it is possible to model the qualitative response of the device in the under-limiting current regime for electrolyte concentrations corresponding to a high degree of double-layer overlap, hence high selectivity. The model is used to predict the low frequency (Warburg-branch) of the impedance response of the system and yields increasing slope depression with increasing DC bias, corresponding to a larger value of the net EOF, as shown in the figure below.

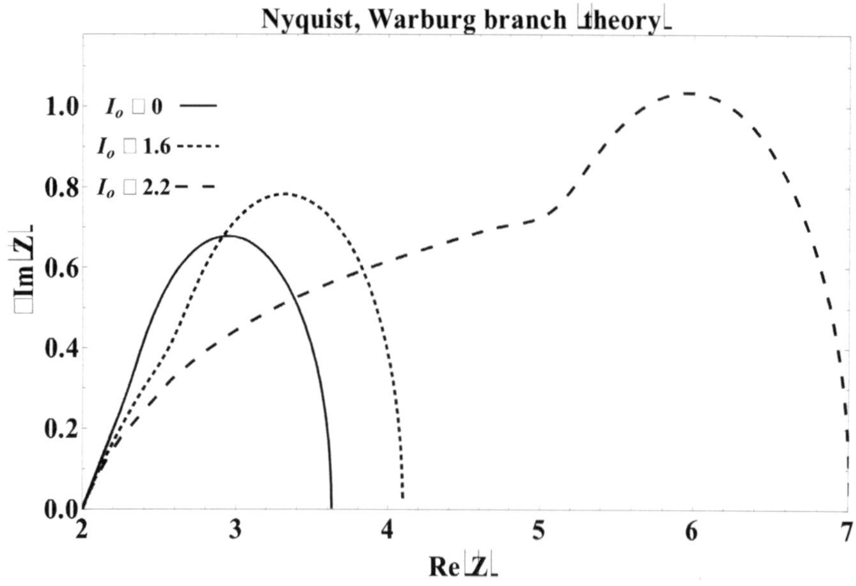

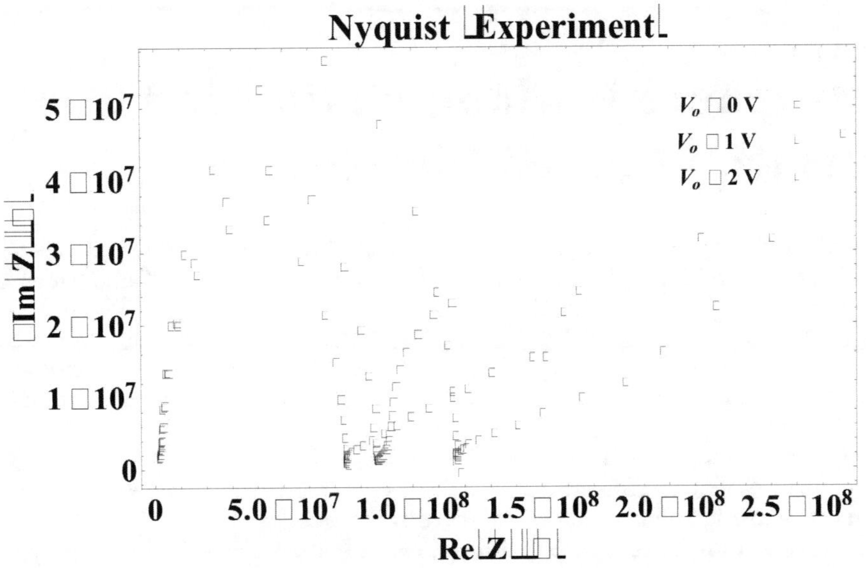

The potentiostatic impedance response of a glass micro-nanochannel device was tested using a 10-4 M KCL electrolyte at three different voltage biases, 0V, 1V, and 2V, corresponding roughly to the same response regimes, relative to the limiting current condition, of (dimensionless) galvanostatic biases in the theory. The experimental impedance spectra contain the bulk-electrode semi-circular branch at high frequencies (right lobe in in the figure above,) as well as a lower-frequency Warburg branch. From the data obtained thus far, the Warburg response appears to be of the finite-length type, exhibiting a maximum in Im(Z) corresponding to the characteristic diffusive relaxation frequency. Because this frequency is below 1 mHz, it has been difficult to reach experimentally without experiencing appreciable drift in the system due to an accumulating concentrate gradient between the device reservoirs and evaporating solution. However, it is clear from the figure that the linear edge of the Warburg branch exhibits the slope-depression effect with increasing voltage. The evident shift of the Warburg/bulk intersection with applied voltage has been observed consistently across multiple runs, all confirming the slope depression. This voltage shift is believed to be the result of electrode reactions, and further tests are being conducted to verify this.

(297b) Dielectrophoretic Manipulation and Controlled Release of Surfactant Based Micelles

Victoria Goodrich, Chemical and Biomolecular Engineering, University of Notre Dame, Notre Dame, IN and Yingxi Elaine Zhu, Department of Chemical and Biomolecular Engineering, University of Notre Dame, Notre Dame, IN

Lipid and surfactant based biocolloids have been of increasing interest as delivery systems for controlled release, targeting, or medical diagnostics via AC-electrokinetic and other techniques. In this work, we examine the behavior of surfactant based micelles under dielectrophoresis (DEP) as a means to rapidly assemble micellar complexes with controlled release of varied encapsulatess. The dynamic response of sodium tetradecyl sulfate (STS) and sodium dodecyl sulfate (SDS) micelles, each 3-5 nm in diameter, is accomplished through single-molecule resolution spectroscopy, namely fluorescence correlation spectroscopy (FCS). We focus specifically on the AC-induced concentration changes of micelles due to applied field of AC-field frequencies of 5 kHz – 20 MHz at an amplitude of 20 V (peak-to-peak). We observe a strong dependence of micelle concentration on ac-field frequency in the region of high field strength. Surprisingly, we also observe an AC-induced dissociation of the micelle structure, resulting in the release of the encapsulated chemicals at low AC-field frequencies (< 10 kHz). In addition, the micellar dissociation is found to be linked specifically to the surface charge of the interior encapsulate.

(297c) Trapping of Nanoparticles with Dielectrophoretic Nano-Probes

Nicholas R. Wood[1], Amanda I. Wolsiefer[1], Robert W. Cohn[2] and Stuart J. Williams[1], (1)Department of Mechanical Engineering, University of Louisville, Louisville, KY, (2)ElectroOptics Research Institute and Nanotechnology Center, University of Louisville, Louisville, KY

Silver gallium nano-probes have been used as dielectrophoretic electrodes to capture sub-micron colloids. The silver gallium nano-probes were fabricated using a unique self-assembly method. The nano-probes are crystalline structures that self-assemble at room temperature and pressure when thin films of elemental silver are introduced to micro-spheres of elemental gallium. The direction of the nano-probe assembly can be controlled by the geometry of the substrate on which they are grown. The diameters of the nano-probes were measured to be approximately 100-500 nm. Using nano-probes and ITO coverslips as electrodes, sub-micron colloids suspended in DI water and aqueous KCl were attracted and repelled using positive and negative dielectrophoretic forces at voltages of less than 1 V_{RMS}. Positive DEP forces successfully captured 200 nm fluorescent polystyrene particles as well as gold colloidal quantum dots as small as 5 nm in diameter. Negative DEP forces were observed acting on fluorescent sub-micron polystyrene particles in solutions of various conductivity levels.

(297d) Coupling Dielectrophoresis At Nano-Constrictions to Concentration Polarization Effects for Enhanced Protein Pre-Concentration

Mikiyas Tsegaye[1], Kuo-Tang Liao[2], Chiafu Chou[3] and Nathan Swami[2], (1)Electrical & Computer Engineering, University of Virginia, Charlottesville, VA, (2)Electrical & Computer Engineering Department, University of Virginia, Charlottesville, VA, (3)Academia Sinica, Taipei, Taiwan

The challenge of sensing rare numbers of biomarkers against a background of high concentration of other matrix proteins within physiologically relevant media requires methodologies for selective pre-concentration of the biomarker in proximity of the sensor. Antibody affinity based chemical depletion methods are unable to achieve the necessary degree of pre-concentration, since the biomarkers are present at billion-fold lower levels than the background proteins in blood [1]. Hence, there is great interest in applying electrokinetic methods, especially within nanofluidic devices, where million-fold levels of pre-concentration have been reported using concentration polarization effects to cause ion exclusion-enrichment due to electrical double layer overlap at the micro-to-nanofluidic interface [2]. However, these methods require several tens of minutes to achieve this pre-concentration and they cannot effectively separate target biomarker proteins against a background of similarly charged and sized proteins in samples of physiological fluids. Dielectrophoresis enables highly selective trapping of bio-particles based on the characteristic frequency response of the dielectric permittivity of the bio-particle versus that of the medium, and can be applied in principle towards selective pre-concentration based on differing dielectric frequency response. However, its application to trapping nanoscale bio-particles, such as ss-DNA and proteins, requires methods to enhance the local field to offset the steep fall in dielectrophoretic trapping forces with particle size [3]. Furthermore, within physiological media of high conductivity, dielectrophoretic trapping is in competition with electrothermal flow due to Joule heating [4]. Herein, we explore the coupling of a DC field offset to negative dielectrophoretic trapping of streptavidin protein biomolecules at nano-constrictions by AC fields (200 V_{pp}/cm, 1 MHz) to enhance the degree of pre-concentration through an electrokinetic force balance [5]. At sub-100 nm constrictions, the addition of a critical DC field offset to the AC field causes the emergence of concentration polarization effects in addition to dielectrophoresis to enable an exponentially enhanced extent of the protein depletion zone across the device, thereby resulting in steep and ultra-fast protein pre-concentration. Using a potential energy diagram across the device to explain protein trapping in terms of energy wells due to energy barriers caused by the constriction and potential profile tilting caused by the DC field, we develop a design parameter to characterize the effectiveness of constrictions for protein trapping based on depth and sharpness of the energy well. We envision that these protein pre-concentration methodologies may be applied towards biomarker discovery, protein crystallization, and rare target sensing for early disease diagnostics.

[1] Polaskova, V.; Kapur, A.; Khan, A.; Molloy, M. P.; Baker, M. S. *Electrophoresis* **2010,** *471-482,* 31

[2] Wang, Y. C.; L., S. A.; Han, J. Million-fold Preconcentration of Proteins and Peptides by Nanofluidic Filter. *Analytical Chemistry* **2005,** 4293.

[3] Swami, N.; Chou, C.-F.; Ramamurthy, V.; Chaurey, V. Enhancing DNA hybridization kinetics through constriction-based. *Lab on a Chip* **2009,** *9,* 3212–3220.

[4] Chaurey, V.; Polanco, C. F.; Chou, C.-F.; Swami, N. S. Floating electrode enhanced constriction dielectrophoresis for biomolecular trapping in physiological media of high conductivity. *Biomicrofluidics* **2012,** *6,* 012806.

[5] Liao, K.T.; Tsegaye, M.; Chou, C.F.; Swami, N. "Nano-constriction device for rapid protein pre-concentration in physiological media through a balance of electrokinetic forces", *Electrophoresis* (2012) (Accepted)

(297e) Combining Focused Ion Beam Milling and Optical Lithography to Fabricate Microfluidic Devices for DNA Dielectrophoresis

Fernanda Camacho-Alanis, Lin Gan and Alexandra Ros, Department of Chemistry and Biochemistry, Arizona State University, Tempe, AZ

Dielectrophoresis (DEP) refers to the migration of particles in a non-uniform electric field due to the interaction of the particle's dipole and spatial electric field gradient. DEP thus finds application in concentration and fractionation. In the case of DC conditions, DEP forces on biomolecules compete with electrokinesis resulting from the combination of electroosmotic flow (EOF) and electrophoresis (EP). In order to achieve trapping of biomolecules in a microfluidic device, the DEP force must overcome EOF and EP. Otherwise, when electrokinesis dominates over DEP, streaming behavior arises. Therefore, the design of the microfluidic device has to focus on the enhancement of the electric field gradient in order to minimize the additional forces. A novel way to create electric field gradients for DEP is using insulator posts inside the microfluidic device well known as i-DEP. This method generates homogenous electric field gradients over the entire depth of a microchannel. In addition, lithography steps are simplified since a single material is used in the fabrication process. Here, we tailored insulating post geometries inside microfluidic channels using focused ion beam milling (FIBM) in combination with optical lithography to improve i-DEP of lambda-DNA over other electrokinetic effects.

Micro-posts in the microfluidic channel were patterned by standard photolithography followed by FIBM of nano-posts in between the micro-posts. A PDMS mold was fabricated from this master and the resultant mold was bonded on a glass slide. The experiments were performed by recording the movements of particles with fluorescence microscopy while applying a DC potential in the channel. As a control experiment, we used a microfluidic device without nano-posts. Finally, numerical simulations using a convection-diffusion model were performed.

Experimental results indicate that only streaming behavior is obtained in the case of micro-posts. However, lambda-DNA was trapped by iDEP when posts fabricated with FIBM are present in the channel. According with numerical simulations, the electric field gradients are enhanced by 2.5 orders of magnitude when nano-posts are integrated between the micro-posts. This study indicates that the novel fabrication process has the potential to improve applications for dielectrophoretic separation, concentration and fractionation of biomolecules.

(297f) Rectification and Rectification Inversion of Ion Currents in Conical Nanopores

Yu Yan and Hsueh-Chia Chang, Chemical & Biomolecular Engineering, University of Notre Dame, Notre Dame, IN

As shown in some recent experiments, ion transport in perm-selective media can exhibit rectification behavior like semiconductor diodes. Such rectification can be a result of asymmetrical geometry (Siwy et al, Europhys Lett 2002), concentration gradient (Cheng and Guo, Nano Lett 2007) and non-uniform surface charge (Daiguji, Oka and Shirono, Nano Lett 2005). Here, we focus on mechanism behind rectification induced by conical geometry and explain why the rectification can reverse with change in bulk concentration, as observed by some recent numerical simulations (Momotenko and Girault, JACS 2011) and experiments (Gilad Yossifon--unpublished data). The conic geometry allows us to carry out rigorous cross-section averaging to produce a one-dimensional Nernst-Planck equation for ion transport of a symmetric electrolyte. The averaged transport equation is complemented by Donnan type boundary conditions due to Boltzmann equilibrium between the boundary ions with the bulk. Using an expansion in a small parameter, corresponding to the ratio of the effective surface charge concentration over the bulk ionic strength, we show that the selectivity jump along the conic pore produces non-equilibrium intra-pore enrichment and depletion under positive/reverse bias and is hence responsible for the rectification phenomenon. We also offer a simple expression for the rectification factor, as a function of geometric parameters, surface charge, voltage and bulk ionic strength, that satisfactorily collapse literature data and data from our collaborators at high bulk ionic strengths. However, at very low bulk concentrations, we find rectification inversion can happen with the increase of applied voltage due to external depletion, which is also asymmetric because of the different diffusion lengths outside the two entrances. By matching the intrapore current with the limiting current due to external depletion (Yossifon et al, PRE 2009), we obtain a simple estimate of the crossover voltage for rectification inversion, which agrees with our numerical simulation. Conic pore rectification is hence due to asymmetric enrichment/depletion both within and outside the nanopore.

(297g) Accurate Predictions of Dielectrophoretic Force and Torque On Many Particles with Strong Mutual Field, Particle, and Wall Interactions

Qianlong Liu, Univ. South Carolina, Columbia, SC and Kenneth Reifsnider, Univ. South Carolina

The basis of dielectrophoresis (DEP) is the prediction of the force and torque on particles. The classical approach to the prediction is based on the effective moment method, which, however, is an approximate approach, assumes infinitesimal particles, and totally neglects the perturbation of the resultant electric field by the particles. Therefore, it is well-known that for finite-sized particles, the DEP approximation is inaccurate as the mutual field-particle, particle-particle, particle-wall interactions become strong, a situation presently attracting extensive research for practical significant applications. In the present paper, we provide accurate calculations of the force and torque on the particles from first principles, by directly resolving the local geometry and properties and accurately accounting for the mutual interactions for finite-sized particles with both dielectric polarization and conduction in a sinusoidally steady-state electric field. Since the approach has a significant advantage, compared to other numerical methods, to efficiently simulate many closely packed particles (one million 2D particles and one hundred thousand 3D particles), it provide an important, unique, and accurate technique to investigate complex DEP phenomena, for example heterogeneous mixtures containing particle chains, nanoparticle assembly, multilayered materials, non-spherical effects, etc.

(297h) Correction for Sample Conductance in the Measurement of the Zeta Potentials of Porous Samples by the Rotating Disk Technique

Paul Sides[1], Kong M Wong[1] and Dennis C. Prieve[2], (1)Chemical Engineering, Carnegie Mellon University, Pittsburgh, PA, (2)Dept. of Chemical Engineering, Carnegie Mellon University, Pittsburgh, PA

The zeta potentials of porous materials are important quantities for post chemical mechanical planarization processes in which highly porous brushes clean particles and debris from wafers after planarization. It is particularly important to know the zeta potential in order to design the brush material to repel particles removed from the wafers. If the particles electrostatically adsorb on the brush material, the effectiveness of the brush declines and the potential for scratching the wafer increases.

The use of the rotating disk geometry for determining the zeta potential of porous materials is the topic of this contribution. The zeta potential of a non-porous material is determined from streaming potential measurements acquired near a planar disk sample affixed to the end of a spindle and rotated. The theory underlying the method makes use of the well established hydrodynamics of the rotating disk.[1]

While the rotating disk approach was originally developed for determination of the zeta potential of planar disk-shaped samples, it has been found that an amplified streaming potential is measured when the sample is porous. For example, one measures several mV of streaming potential in the vicinity of a disk-shaped porous quartz frit rotating on its axis in 1 mM electrolyte. Naïve application of the theory developed for planar surfaces gives absurdly large apparent zeta potentials. A new theory of the streaming potential caused by a rotating porous body,[2] based on hydrodynamics developed for a rotating filter,[3] was applied to the experimental results. The exercise showed that the large streaming potentials were clearly related to the porosity of the sample; a predicted dependence of the streaming potential on the square of the rotation rate was confirmed. It was found that the permeability of the quartz frit could be found, but the magnitudes of the zeta potential determined from the theory were now much too small.

The shortcoming of the new theory was neglect of the sample conductance. The sample provides a channel for leakage of streaming current from the periphery backwards, much like surface conductivity provides a path for streaming current to flow back to its origin without contributing to the streaming potential. Analytical and numerical analysis of the streaming potential near a rotating porous sample of finite thickness reveals that two parameters control the short-circuiting. The first is the ratio of the sample thickness to the sample radius. The second parameter is the ratio of the conductance of the porous layer to the conductance of the electrolyte, given by $s_p h/s_e a$, where s is ionic conductivity, p denotes the porous sample, e denotes the liquid electrolyte, h is the sample thickness, and a is the sample radius. When the dimensionless sample thickness is less than 0.2 and the conductivity parameter is less than 0.01, the short-circuiting of the streaming current is small and the analytical solution provides a good result. Larger values of these parameters mean that correction for leakage of streaming current leakage is necessary. For example, the streaming potential measured in the vicinity of a 25 mm diameter porous quartz frit (porosity = 0.5) that is 5 mm thick is only about 50% of what is expected if the short circuiting effect is neglected.

1. P. Sides, J. Newman, J. Hoggard, D. Prieve, "Calculation of the Streaming Potential Near a Rotating Disk," *Langmuir*, **(2006) 22** 9765-9769.
2. P. Sides, S. Mukka, D. Prieve, "Measurement of the zeta potential of open porous structures by means of a sample rotating on its axis," AIChE Annual Meeting 2011, Minneapolis, MN. 18OCT2011.
3. Joseph, D. D., "Note on Steady Flow Induced by Rotation of a Naturally Permeable Disk," *Quart. J. Mech. Appl. Math.* **(1965) 18** 325-331.

(297i) High Pressure-to-Voltage Energy Conversion Efficiency in Nanofluidic Channels

Dirk Gillespie, Molecular Biophysics and Physiology, Rush University Medical Center, Chicago, IL

The electrokinetic behavior of nanofluidic devices is dominated by the electrical double layers at the device walls. In many contemporary nanofluidic devices, the surface charge of the channel walls is large enough that the nonlinear coupling between the "simple" interactions between the wall and ions gives the device novel properties that change with ion type and concentration. The simplest nanofluidic device geometry is the channel with two charged walls separated by ~100 nm. Here, it is proposed that the layering of large ions at the wall/liquid interface of nanofluidic channels (due to wall-ion and ion-ion correlations) can be used to achieve high efficiency (possibly >50%) in the conversion of hydrostatic energy into electrical power. Large ions tend to produce peaks and troughs in their concentration profiles at charged walls, producing high concentrations far from the walls where the ions' pressure-driven velocity is high. This increases the streaming conductance and the energy conversion efficiency. The results are computed with density functional theory of fluids, a theory that includes ion-ion correlations in a way this both computationally efficient and thermodynamically accurate.

3D Carbon Cages For Dielectrophoretic-Based Bioparticle Separation/Concentration Applications

Abstract ID# 275627

*Víctor H. Pérez-González,[1,2] Vinh Ho,[2] Lawrence Kulinsky,[2] Blanca H. Lapizco-Encinas,[3] Sergio O. Martínez-Chapa,[1] Marc Madou,[2,4]**

1 BioMEMS Research Chair, Tecnológico de Monterrey, Monterrey, NL, México.
2 Mechanical & Aerospace Engineering, University of California, Irvine, Irvine, CA, US.
3 Microscale Bioseparations Laboratory and Department of Chemical Engineering, Tennessee Technological University, Cookeville, TN, US.
4 Ulsan National Institute of Science and Technology, World Class University Program, Ulsan 689-79, South Korea.

**Correspondence author. Email: mmadou@uci.edu*

ABSTRACT

Dielectrophoresis (DEP) is an electrokinetic phenomenon defined as the force experienced by polarizable particles due to their interaction with a spatially-non-uniform electric field. The magnitude and direction of this movement depends upon the interaction between an induced electrical dipole and the electric field, as well as on the relation between the dielectric properties of the particles and the suspending medium in which the particles are immersed. Carbon electrode based DEP (CarbonDEP) is a novel approach to microparticle manipulation. 3D Carbon structures, used here as electrodes for DEP applications, are produced by a sequence of photolithographic patterning and pyrolysis of organic precursors. Carbon has a wider electrochemical stability window than most materials employed in metal-electrode based DEP. Carbon is also biocompatible, making it attractive for biotechnology applications. This work presents new developments in the area of fabrication of CarbonDEP-based microfluidic platforms. Polypyrrole (PPy) posts were fabricated over planar gold leads covered with PPy. The posts were then connected on the top by a layer of laterally grown PPy. PPy cages were tested with polystyrene particles with diameters of 1 μm. It was found that PPy cages allow for a higher trapping efficiency compared to planar electrodes and also to flat electrodes with 3D posts. Subsequent pyrolysis step will convert PPy organic precursor into carbon, thus producing Carbon cages. Simulation work was carried out using COMSOL Multiphysics platform to analyze the electric field distribution within a microchannel where DEP particle separation is taking place. Applications of this work include, but are not limited to, environmental monitoring, food safety control, clinical analysis, and clean energy production.

INTRODUCTION

Environmental monitoring, food safety engineering, and clinical analysis are just a short list of many biotechnological applications that require the manipulation of bioparticles. Separation and isolation of bioparticles are two of the processes that are required during most of biotechnological experiments that involve particles. Although commercial equipment, such as Fluorescence Activated Cell Sorters (FACS) or Magnetic Activated Cell Sorters (MACS) [1, 2], are readily available for performing these tasks, a common FACS equipment is bulky and to work well it needs a large amount of sample. MAC sorters, on the other side, require the attachment of magnetic beads to the particles of interest, therefore modifying the original sample. Dielectrophoresis (DEP) is an electrokinetic (EK) technique that offers the advantages of miniaturization, i.e. it can be implemented on a microfluidic chip. DEP allows for nondestructive manipulation of the sample and it has been used successfully to separate, isolate, displace, and filtrate particles as well as to determine the dielectric properties of microparticles [3].

A non uniform electric field distribution along the microfluidic chip is required to induce DEP in the particles of interest [3]. In order to produce these non uniform fields, metal planar electrodes of different geometries are commonly used. These electrodes are generally placed on the bottom of the microfluidic channel that contains the sample [4]. However, this approach has some limitations, the most important of which is the lack of efficient particle manipulation throughout the bulk of the sample (the dielectrophoretic force will be exerted on the particles in close proximity to the metal electrodes, while particles in the bulk of the solution will pass through unaffected). An alternative to metal electrode based DEP is insulator based DEP (iDEP), where an array of 3D insulating structures is fabricated within the microfluidic channel. An electric signal is then applied between two needle electrodes, which are positioned at the inlet and outlet of the microfluidic chip in order to create an electric field across the channel. The insulating structures modify the distribution of the electric field creating nonuniformities [5, 6]. While iDEP is a fairly inexpensive technique, it has the drawback that the voltage required to induce DEP on the particles sometimes surpasses 1000V, creating Joule heating that can damage the samples and raising the safety of use concerns in laboratories.

Carbon electrode based DEP (Carbon DEP) is a novel approach to microparticle manipulation that brings together the advantages of both, metal-electrode-based DEP and iDEP. In Carbon DEP, an array of carbon posts is fabricated on top the conventional planar electrodes, therefore creating non uniformities in the electric field not only in the vicinity of the microchannel's bottom, but also in the bulk of the fluidic sample. Because each carbon post acts as an electrode, the electric potential required is much lower than that needed for iDEP applications. Also, Carbon is mechanically and chemically stable, biocompatible and has a wider electrochemical window than that of common metals used to fabricate planar electrodes [7].

The work presented in this paper is a novel modification of previously reported CarbonDEP [7, 8]. Through the electropolymerization of polypyrrole [9], a new 3D electrode geometry has been developed. These new electrodes are shaped as cages, where posts grown over a single planar electrode have been connected by the layer of polypyrrole on top. These cages can then be turned into Carbon through pyrolysis. This geometry significantly increases the volume of the sample affected by DEP, hence enhancing the trapping efficiency of the device.

THEORY OF DIELECTROPHORESIS

Dielectrophoresis [10] is an electrokinetic (EK) phenomenon defined as the interaction between an induced dipole and a non uniform electric field. A polarizable particle must be suspended in an electrolytic solution for DEP to take place. DEP can be induced employing Direct Current (DC) or Alternating Current (AC) fields [3, 11-13]. The dielectrophoretic force experienced by a spherical particle is defined as:

$$\mathbf{F}(\omega) = 2\pi\varepsilon_m r^3 \operatorname{Re}[f_{CM}]\nabla E_{rms}^2 \quad (1)$$

where ε_m is the permittivity of the suspending medium, r is the particle radius, and f_{CM} is the Clausius-Mossotti (CM) factor, which is defined as:

$$f_{CM} = \frac{\varepsilon_p^* - \varepsilon_m^*}{\varepsilon_p^* + 2\varepsilon_m^*} = \frac{\sigma_p^* - \sigma_m^*}{\sigma_p^* + 2\sigma_m^*} \quad (2)$$

where $\varepsilon^* = \varepsilon - j\sigma/\omega$ and $\sigma^* = \sigma + j\varepsilon\omega$ refer to the complex permittivity and complex conductivity respectively. In (2), subscripts p and m refer to particle and medium respectively, and ω is the angular frequency of the applied electric signal [14]. The CM factor is a measure of the ease of polarizability of a particle in a given suspending medium. The dielectrophoretic force can be positive, pulling particles towards the zones of highest electric-field-squared gradient (positive CM factor), or negative, repelling particles from such zones (negative CM factor).

COMPUTATIONAL SIMULATION

A computational model was built with COMSOL Multiphysics 4.3a © (COMSOL Inc., Burlington, MA) to study the distribution of the electric field in the microfluidic chip. COMSOL Multiphysics is a computational tool based in the Finite Element Method (FEM), which is capable of simulating physical phenomena for which no closed form analytical solution exists.

Since each planar electrode contains 30 merged posts, the geometry employed to perform the computational study can be simplified to a basic building block that consists of two planar electrodes, each with two merged posts. Such geometry is shown in Figure 1 where the planar interdigitated electrodes are 35µm wide; the posts have a diameter of 25µm and a height of 24µm; the center to center distance between posts is 75µm.

The Electric Current Module was employed to analyze the model in COMSOL Multiphysics. Boundary conditions of electric potential and ground were employed in the electrodes to run a stationary analysis. The geometry was enclosed in a cube with a surface defined as an electric insulator to account for the microfluidic channel walls. These boundary conditions are needed to solve Laplace's equation. The mesh consisted of 550,221 tetrahedral elements.

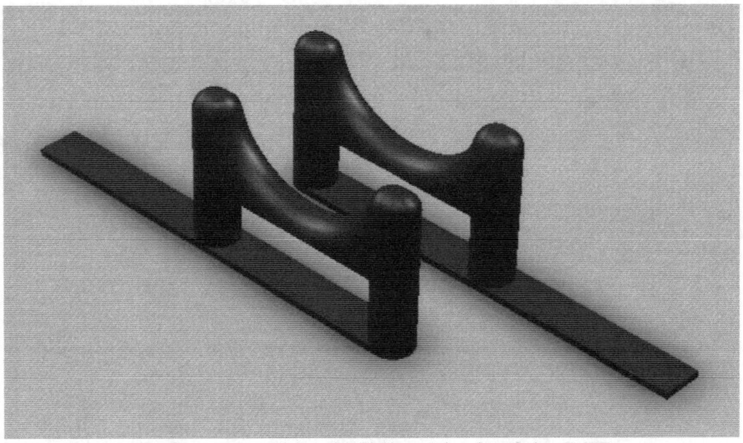

Figure 1. Representative building block of the DEP cages.

EXPERIMENTAL SETUP

A DEP device with the geometry described above was fabricated following the process shown in Figure 2 at the Integrated Nano Research Facility (INRF) and the BioMEMS Research Laboratory at the University of California, Irvine. A 4'' single-crystal Si (001) wafer with a 40nm thick thermally grown silicon dioxide layer was used as the substrate material. A 20nm thick chromium (Cr) adhesion layer and a 60nm thick gold (Au) layer were deposited via e-beam evaporation. These layers were then patterned and etched using standard lithography techniques to define the base interdigitated electrodes over which the DEP cages were grown. Shipley's 1808 photoresist was spin coated on the wafer in two steps: at 500rpm for 10 seconds and then at 3500rpm for 30 seconds. The photoresist was soft baked for 60 seconds at 110°C. After soft baking, it was exposed to UV light (100mJ/cm^2 dose) with the Karl Suss MA6 Aligner Exposure system (SUSS MicroTec, Garching, Germany). For development of the exposed photoresist, Microposit MF-319 Developer (Rohm and Hass Electronic Materials, Marlborough, MA) was employed. Au and Cr etchants (Sigma-Aldrich, St. Louis, MO) were used to etch the exposed metals. Acetone and isopropanol were sprayed over the wafer to remove the photoresist.

A polypyrrole (PPy) layer was grown over the Au interdigitated electrodes employing a solution of 0.1M pyrrole monomers (Sigma-Aldrich, St. Louis, MO) and 0.1M NaDBS (sodium dodecylbenzenesulfonate) (Sigma-Aldrich, St. Louis, MO) under a bias of 0.65V vs Ag/AgCl reference electrode (Fisher Scientific Corp., Waltham, MA) for 3 minutes, a platinum wire was used as counter electrode. Control of the polypyrrole deposition was carried out using a PC4/750 Potentiostat (Gamry Instruments, Warminster, PA).

AZ P4620 (AZ Electronic Materials, Branchburg, NJ) photoresist was spun at 2000rpm for 40 seconds to produce a 12μm thick layer over the PPy layer. The resist was soft-baked at 110°C for 80 seconds. A second 12μm thick layer was deposited on top of the first one. Soft bake was performed at 115°C for 3 minutes. UV exposure through a mask was performed with the Karl Suss MA6 Aligner Exposure system (1950mJ/cm^2 dose). To develop the photoresist AZ400K developer was used in 1:4 ratios with DI water. An array of cylindrical wells (24μm deep with a diameter of 25μm) was obtained after photoresist development.

To grow PPy posts the same process previously mentioned was followed. The deposition rate of PPy under these conditions is of 1μm/min. The PPy was electrodeposited for 22 minutes to grow the posts as shown in Figure 2. Since the base structure over which the posts were grown are interdigitated electrodes, it is very important to prevent a short circuit between the adjacent fingers. Without removing the photoresist mold used to grow the posts, another 24μm photoresist layer was deposited and patterned (following the process mentioned above) to create wells over each finger of the array. The final PPy layer was then grown, the deposition time required to merge the posts was 11 minutes. Finally, a step of photoresist removal was performed by rinsing the device with acetone and isopropanol.

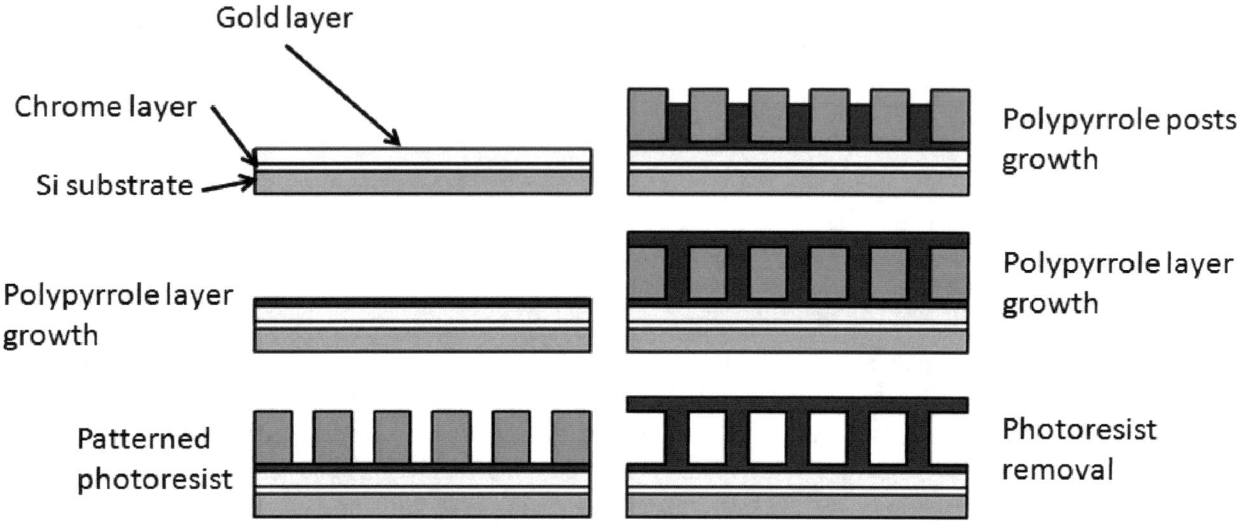

Figure 2. Process flow for the fabrication of polypyrrole electrode cages.

The microfluidic components were fabricated using double-sided pressure-sensitive adhesive and plastic. An image of the fully assembled microfluidic chip is shown in Figure 3.

DI water with a conductivity of 0.5μS/cm was employed as suspending medium. Conductivity was measured using a bench conductivity meter CON510 (Oakton Instruments, Vernon Hills, IL). Polystyrene beads with a diameter of 1μm color dark red (Fluka Biochemika, Sigma-Aldrich, St. Louis, MO) were employed in this work. The sample containing the microspheres was introduced to the microdevice employing a micropipette and reservoirs were then sealed to avoid pressure gradients affecting particle movement inside the microchannel. The microfluidic chip was mounted under an Eclipse LV100 optical microscope (Nikon Instruments Inc., Melville, NY) with an attached SPOT RT KE CCD camera (Diagnostic Instruments Inc., Sterling Heights, MI). Electric signal was provided to the microdevice by a synthesized function generator DS345 (Stanford Research Systems, Sunnyvale, CA) with frequency range from 1Hz to 33MHz and output voltage up to 20Vpp. The various dielectrophoretic responses on the chips with different electrode geometries were video recorded.

Figure 3. Microfluidic chip. Gold planar electrodes covered by a thin layer of PPy. Microfluidic channel and chamber fabricated with pressure-sensitive double-sided adhesive and plastic.

RESULTS

COMSOL Multiphysics was used to perform the computational simulations. Figure 4 shows the distribution of the gradient of the squared rms electric field, $\nabla {E_{rms}}^2$, in a microfluidic device with electrode geometry shown in Figure 1. The distribution has been plotted in a logarithmic scale in order to provide an order of magnitude approximation. Figure 4a shows a top view of the device. It can be seen that the dielectrophoretic force is biggest at the posts' edges. Figure 4b demonstrates the side view of the microfluidic chip where post electrodes support the top PPy layer. It can be observed how the largest field gradients are located at the edges of the bottom electrode leads and also at the edges of the merged top layer.

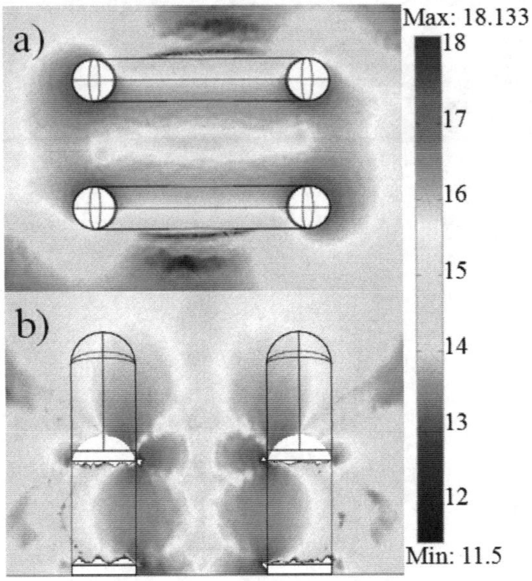

Figure 4. COMSOL modeling of the gradient of squared rms electric field $\nabla {E_{rms}}^2$: (a) top view; (b) side view.

Computational modeling is important for guidance towards more efficient 3D electrode geometries for DEP applications. For example, Figure 4b demonstrates that having a merged layer atop of posts it beneficial in spreading the high field throughout the bulk of the microfluidic channel.

Experimental work was carried out in these devices to verify the information gathered from computational modeling. A sequence of video frames is shown in Figure 5. Figures 5a-b present negative DEP induced by a set of planar interdigitated electrodes. Figures 5c-d show negative DEP induced by a set of planar interdigitated electrodes with PPy posts. The frequency of the AC electric potential externally applied to the electrodes in was 1MHz with the "peak-to-peak" amplitude of 8V (Figure 5b) and 10V (Figure 5d).

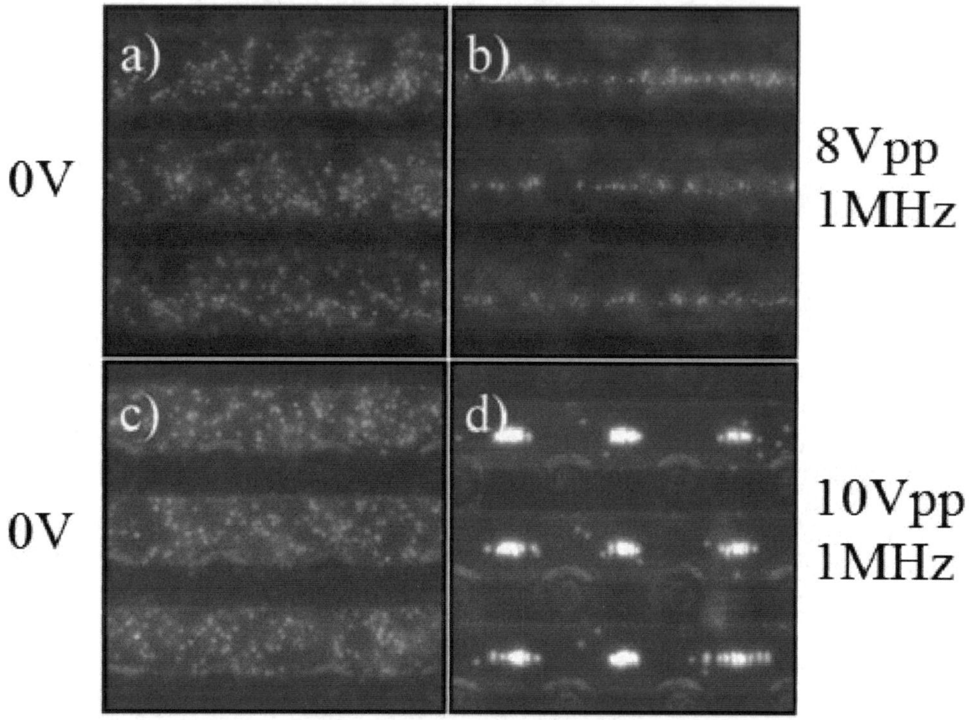

Figure 5. Sequence of 1μm polystyrene beads movement due to negative DEP under an AC potential: images a) and b) correspond to planar interdigitated electrodes, c) and d) to planar interdigitated electrodes with PPy posts. Images a) and c) present initial states (no excitation signal) while b) and d) show the DEP response after 1 minute of electric stimulation (for planar electrodes: 8Vpp at 1MHz; for 3D electrodes: 10Vpp at 1MHz).

Figure 5b shows an increased (compared to Figure 5a) concentration of beads in between interdigitated fingers. Since this is a negative DEP, many particles are pushed out of focal plane to the top and bottom of the channel and we can see many of these out-of-focus particles as blurry images. On the other hand, Figure 5d (representing negative DEP on 3D electrode cages) shows very clear trapping zones with very few particles moving out-of-focus.

CONLUSIONS

A new approach for enhancing dielectrophoretic particle trapping efficiency has been presented. Preliminary results demonstrate a significant promise of this novel technology. Future work will quantify particle trapping efficiency; comparison of the trapping efficiencies of 3D PPy electrodes with 3D carbon electrodes will be undertaken as well.

REFERENCES

[1] Nolan, G. P., Fiering, S., Nicolas, J. F., Herzenberg, L. A., *Proceedings of the National Academy of Sciences* 1988, *85*, 2603-2607.

[2] Miltenyi, S., Müller, W., Weichel, W., Radbruch, A., *Cytometry* 1990, *11*, 231-238.

[3] Morgan, H., Green, N. G., *AC electrokinetics : colloids and nanoparticles*, Research Studies Press, Baldock, Hertfordshire, England 2003.

[4] Ermolina, I., Morgan, H., *Journal of Colloid and Interface Science* 2005, *285*, 419–428.

[5] Moncada-Hernandez, H., Baylon-Cardiel, J. L., Pérez-González, V. H., Lapizco-Encinas, B. H., *Electrophoresis* 2011, *32*, 2502-2511.

[6] Gallo-Villanueva, R. C., Pérez-González, V. H., Davalos, R. V., Lapizco-Encinas, B. H., *Electrophoresis* 2011, *32*, 2456-2465.

[7] Martinez-Duarte, R., Renaud, P., Madou, M. J., *Electrophoresis* 2011, *32*, 2385-2392.

[8] Martinez-Duarte, R., Gorkin-Iii, R. A., Abi-Samra, K., Madou, M. J., *Lab on a Chip* 2010, *10*, 1030-1043.

[9] Ho, V., Shimada, M., Szeto, D., Mukherjee, P. P., *et al.*, 2010, 76472F-76472F.

[10] Pohl, H. A., *Journal of Applied Physics* 1951, *22*, 869-871.

[11] Jones, T. B., *Electromechanics of particles*, Cambridge University Press, Cambridge ; New York 1995.

[12] Green, N. G., Ramos, A., Morgan, H., *Journal of Physics D: Applied Physics* 2000, *33*, 632-641.

[13] Jones, T. B., *IEEE Engineering in Medicine and Biology Magazine* 2003, *22*, 33-42.

[14] Jones, T. B., *Electromechanics of Particles*, Cambridge University Press, USA 1995.

(334b) Particle and Molecular Species Separation Via Continuous Electroosmotic Flow with Orthogonal Optical Pressure

Sarah J.R. Staton[1], Alex Terray[2], Greg Collins[3] and Sean Hart[3], (1)National Research Council, Washington, DC, (2)Chemistry & Physics, Naval Research Laboratory, Washington D. C., (3)Chemistry & Physics, Naval Research Laboratory, Washinton D. C., DC

The separation of particles and molecular species is an important first step for complex biological or environmental samples in preparation for analysis by traditional analytical techniques. It is also desirable that this preparation occur in a continuous high-throughput manner without direct manipulation to decrease chances of contamination while maintaining production. A new promising application of electroosmotically induced "plug" flow in combination with an orthogonal optical force on a microfluidic platform was developed. This method selectively removes particle matter from a mixed sample stream of molecular species and variously sized particles. The use of electroosmotic flow reduces cross-sectional velocity heterogeneity while increasing the addressability of laminar layers via optical force. Both experimental and theoretical studies have been performed to evaluate the ability and the accuracy of the method to remove particles from the sample stream while minimizing the effects of diffusion and dilution on the remaining molecular stream. Various sized and refractive indexed particles were combined with a water soluble dye and introduced to a pinched flow microfluidic channel. Along the length of the channel a mildly focused 1064 nm Nd YAG laser applied orthogonally to the fluid flow elicited selective particle movement for all particle types as well as combinations of two particles types with the dye. Theoretical calculations using COMSOL, a finite element multiphysics model, predicted the diffusive losses of the molecular species at less than five percent. This technology along with other applications of optical chromatography allow for complex sample stream separation and purification.

(334c) Modulating Electric Fields At Patterned Collectors for the Alignment of Sub-100nm Electrospun Nanofibers

Yi-Hsuan Su[1], Vasudha Chaurey[2], Frank Block[3] and Nathan Swami[2], (1)Electrical & Computer Engineering, University of Virginia, Charlottesville, VA, (2)Electrical & Computer Engineering Department, University of Virginia, Charlottesville, VA, (3)University of Virginia, Charlottesville, VA

Aligned nanofiber scaffolds are commonly applied within a variety of applications that require highly directional cues, such as tissue engineering, chemical sensing and ferroelectrics. In many cases, scaffolds composed of nanofibers with successively smaller sizes are required. Nanofiber alignment is usually enabled by rotating mandrel-based mechanical approaches [1] or by electrical approaches for alignment based on field modification using insulator gaps [2] or dielectrics patterned on the collector [3]. However, the alignment of successively smaller-sized electrospun nanofibers to a high degree is challenging, since smaller fibers experience larger distortions arising from their greater flux and their longer time within the instability region. Furthermore, they are more easily disrupted under shear forces due to alignment using mandrel-based mechanical approaches. Herein, we explore the optimization of electrical methods to enhance the effectiveness of alignment of sub-100 nm fibers through methodologies to vary the transverse field along the nanofiber axis and repulsive fields between neighboring nanofibers due to their residual charge [4]. In this manner, we present a quantitative design methodology for optimizing collector properties, such as gap width, material and permittivity to optimize the spatial extent of the transverse fields, as well as charge density and dielectric properties of the electrospun material to optimize repulsive fields due to residual charge. The application of nanofiber scaffolds aligned through these methods presents a tremendous potential for nerve tissue regeneration [5].

[1] Matthews, J. A.; Wnek, G. E.; Simpson, D. G.; Bowlin, G. L. Electrospinning of collagen nanofibers. Biomacromolecules 2002, 3, 232.

[2] Li, D.; Wang, Y.; Xia, Y. Electrospinning nanofibers as uniaxially aligned arrays and layer-by-layer stacked films. Adv. Mater. 2004, 16, 361.

[3] Yan, H.; Liu, L.; Zhang, Z. Alignment of electrospun nanofibers using dielectric materials. *Applied Physics Letters* **2009,** *95* (14), 143114-1431-3.

[4] V. Chaurey, P. Chiang, C. Polanco, R. Su, C.F. Chou, N. Swami. "Interplay of electrical forces for alignment of sub-100 nm electrospun nanofibers at insulator gap collectors", *Langmuir* (2010), 26 (24), pp 19022–19026.

[5] Neal, R.; Tholpady, S.; Foley, P.; Swami, N.; Ogle, R.; Botchwey, E.; "Alignment and composition of laminin-polycaprolactone nanofiber blends enhance peripheral nerve regeneration", *Journal of Biomedical Materials Research: Part A* (2011), 100, 406-423.

(334d) A Microfluidic Platform for Impedance Analysis and Characterization of Human Umbilical Vein Endothelial Cells

Vanessa Velasco, Mechanical Engineering, University of Louisville, Lousiville , KY, Mark Gruenthal, Mechanical Engineering, University of Louisville , Louisville, KY and Stuart J. Williams, Department of Mechanical Engineering, University of Louisville, Louisville, KY; Birck Nanotechnology Center, Purdue University, West Lafayette, IN

The characterization of endothelium morphology and permeability under fluid shear stress can provide essential information regarding the onset of different pathological conditions, as well as, the uptake of drugs and biomolecules. Real-time assessment of the integrity of cell monolayers and cell motility has been accomplished by implementing electrical impedance analysis. In this study, we report a micro-electrode array biosensor incorporated into a microfluidic platform for the impedance spectroscopy of Human Umbilical Vein Endothelial Cells (HUVECs) under physiological fluid shear stress. The biosensor consists of two adjacent and identical microfluidic channels which allows for simultaneous assessment of impedance spectroscopy of a HUVEC monolayer and that of the cell culture media alone. Additionally, the biosensor is attached to a custom designed electronic circuitry that was built for time-dependent multiplexed measurements for the different electrodes. Flow induced changes in HUVEC morphology, permeability, and relevant electrical parameters were extracted from the impedance spectroscopy. This biosensor can be used as an *in vitro* model for flow induced changes in cell morphology.

Magnetic Capture of Melanoma Cells from Whole Blood

Zhixi Qian[a], *Paul W. Todd*[b], *and Thomas R. Hanley*[a]
[a]*Department of Chemical Engineering, Auburn University, Auburn, AL USA*
[b]*Techshot, Inc., 7200 Highway 150, Greenville, IN USA*

Recent advances in magnetic cell selection have enhanced the possibility of isolating cells of interest. The detection of fetal cells in maternal blood, circulating stem cells and circulating tumor cells will soon be a technology that can be transferred to point-of-care (POC) diagnosis. As a component of a POC portable rare cell analyzer a three-stage magnetic trap for magnetically labeled rare cells was designed and tested. The trap design consists of a flow channel with three bends and a permanent magnet positioned at the outer curve of each bend. Findings are reported on the computational fluid dynamic analysis of the design, testing with model commercial magnetic beads used in cell labeling, and testing with immunomagnetically labeled cultured melanoma cells in the presence of whole blood.

To effectively employ cancer detection by point-of-care devices, the coverage of the general population will need to expand considerably. Local clinics and neighborhood health centers would be effective sites for early cancer indicators. Local clinics, however, lack both the complex and expensive analytical devices as well as qualified technicians. To expand early detection, testing would have to be inexpensive and well accepted by the population targeted for screening.

A mature circulating tumor cell (CTC) detection system would require a biomarkers, such as antibodies, to mark the targeted CTCs; a separation system using various techniques (magnetic separation, for example) to separate CTCs from other cells and a detection system to identify and count the CTCs in the sample.

The computational results provide cell flow data along stream lines as a function of position in the trap. The measurements of field profiles of the magnets were used to determine forces applied as a function of position in the trap. Testing with 3.27 μm diameter standard magnetic beads (Seragen) indicated that at least 99.998% of these strongly magnetic particles are captured within the trap, most in the first stage. Testing with cultured CRL14777 melanoma cells labeled with anti-CD146 1.5 μm diameter beads indicated that 90 + 10% are captured at the first stage. The cell trap requires no power, fits within a point-of-care kit and has now been found to meet functional requirements for incorporation into a point-of-care circulating tumor cell analysis system.

Magnetic separation technology has received considerable attention offering unique advantages over other techniques. Magnetic separation devices are relatively inexpensive to produce, physically and chemically stable, biocompatible and environmentally safe. In addition, biological samples exhibit virtually no magnetic background, and thus highly sensitive measurements can be performed in turbid or otherwise visually obscured samples without further processing.

Today, researchers have developed many CTC detection methods effective in detecting tumor cells in blood; however, these systems are still inadequate for the application of point-of-care detection. Current methods still require complex operation, and development is needed before the CTC detector is operational with a relatively simple structure applicable for clinical use. Some applicable designs are still not practical in use. A key factor is the separation progress. The flow rate should be low enough to provide adequate time to make the labeling and capturing of CTCs successful, with current devices unable to satisfy the balance between efficiency and accuracy. The cost for both device and operation can be reduced as well.

During the research task CFD software simulation and mathematic methods were applied. Computational fluid dynamic analysis of the channel design was accomplished using the FLUENT TM analysis package, which performs modeling of fluid flow and tracking particles' movement in complex geometries. With the help of ICEM CFDTM, a mesh file of the channel geometry was developed, and a laminar flow model was employed for calculation. For the simulation of magnetic particles, the channel was simplified to a 2-D rectangle shape. A discrete model was applied to simulate the particle movement. Tracking of particles movements was performed to show the particle trapping efficiency.

The trajectories of magnetic particles were simulated by choosing a set of coordinates within the channel and adding the flow and magnetophoretic velocity vectors as a function of position in the mesh. At each step in time the velocity is recalculated to account for simultaneous changes in streamline velocity, dy/dt, and magnetophoretic velocity, dx/dt.

The simulations completed in this study confirm that the proposed magnetic trap design successfully traps positively labeled cells. The majority of flow in the trap travels in a low velocity, indicating a well-developed laminar flow. The stream line illustrated little turbulence in the trap which makes it possible to use a two-dimensional model for the particle simulation. The simulation result showed the magnet trap is able to catch all the magnetic particles with one pass, while the efficiency of capture drops with a higher flow rate; at a flow rate of 3ml/min, the capture efficiency drops to 50%. From the figure it is obvious that, at x above 4 mm, the magnet particle won't be caught. With the current design, the majority of magnetic beads are captured by the first pair of magnets, and the three-stage design ensures the capture of all beads. The magnetophoretic velocity of all beads and cells analyzed to date is much greater than the fluid flow velocity in the magnetic trap and is sufficient to cause cells to be captured by the lowest-strength magnet in the current design. As the magnetic force is strongest at edge of the magnet poles, particle movement is affected with both v_x and v_y. This prediction is corroborated by experimental tests in which strong magnetic beads and labeled cancer cells are shown to be trapped by the weakest (first) magnet in the staged multimagnet series.

Key words: point-of-care, magnetic cell separation, FLUENT, rare cells in blood

(334f) The Crystal Orientation of Aluminium Coatings Deposited From Ionic Liquids

Yong Zheng, State Key Laboratory of Multiphase Complex Systems, Institute of Process Engineering, Chinese Academy of Sciences, Beijing, China, Jianmin Zhang, State Key Laboratory of Multiphase Complex System, Institute of Process Engineering, Chinese Academy of Sciences, Beijing, China, Xingmei Lu, Institute of Process Engineering, Chinese Academy of Sciences, Beijing, China and Suojiang Zhang, 1.State Key Laboratory of Multiphase Complex System, 2.Key Lab of Green Process and Engineering, Institute of Process Engineering, Chinese Academy of Sciences, Beijing, China

The Crystal Orientation of Aluminium Coatings Deposited from Ionic Liquids

Yong Zheng, Jianmin Zhang, Xingmei Lu, Suojiang Zhang*

Beijing Key Laboratory of Ionic Liquids Clean Process, State Key Laboratory of Multiphase Complex Systems, Institute of Process Engineering, Chinese Academy of Sciences, 100190, Beijing, P. R. China

**Corresponding author: sjzhang@home.ipe.ac.cn*

Early in 1950s, Hurley et al. first found that aluminium could be electrodeposited from the ionic liquid consisting of *N*-ethylpyridinium chloride and $AlCl_3$ at low temperature. From then on, extensive and systematical research on the electrodeposition of aluminium from chloroaluminate ionic liquids has been reported in the literature [1,2]. However, more efficient ionic liquids should be developed and investigated.

In this work, Lewis acidic 1-allyl-3-methylimidazolium chloroaluminate ([Amim]Cl/$AlCl_3$) ionic liquids were used as promising electrolytes in the low-temperature electrodeposition of aluminium. Compared with the conventional ionic liquids, [Amim]Cl/$AlCl_3$ based ones have more favorable physicochemical properties. Smooth, dense and well adherent aluminium coatings were electrodeposited from [Amim]Cl/$AlCl_3$ on the substrates at 10-25 mA/cm^2 and 313.15-373.15 K. It was found that the crystallographic orientations of aluminium changed along with the experimental conditions (see Figure 1). The results indicate that the influence of experimental parameters on the crystal growth of aluminium deposits follows the order, temperature > current density > molar ratio of [Amim]Cl/$AlCl_3$. The low temperature and high current density usually increase the peak intensity of (200) plane, thus promote the crystal growth in this direction on the substrates. Furthermore, the changes in crystal patterns can also influence the surface morphology of deposits. According to the measurements of SEM and XRD, the bright, smooth and well adherent deposits almost exhibit a preferred orientation of (200).

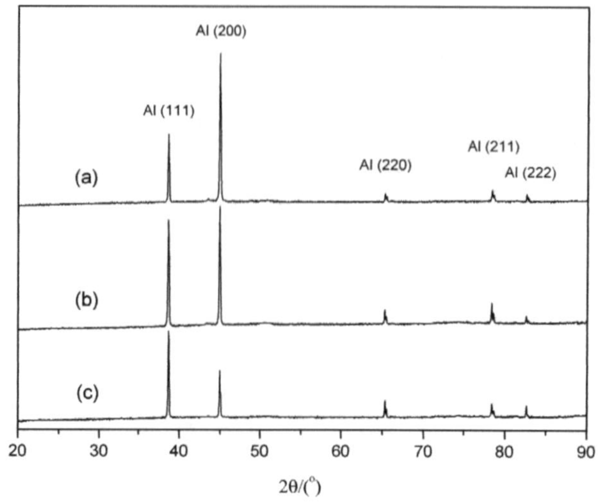

Figure 1 XRD pattern of the aluminium coatings deposited from 1:2.0 [Amim]Cl/$AlCl_3$ at 20 mA/cm^2. The deposition temperature (K): (a) 313.15, (b) 353.15, (c) 373.15.

References:

[1] Liu QX, Zein El Abedin S, Endres F. *Surf Coat Technol*. 2006, 201: 1352-1356.

[2] Yue GK, Zhang SJ, Zhu YL, Lu XM, Li SC, Li ZX. *AIChE J*. 2009, 55: 783-796.

Acknowledgements:

This work was supported financially by the National Basic Research Program of China (2009CB219901), National Key Technology Research and Development Program of the Ministry of Science and Technology of China (2012BAF03B01), National Natural Science Foundation of China (20906096) and Open-end Fund of State Key Laboratory of Multiphase Complex Systems (MPCS-2011-D-06).

(334g) A Nanoporous Membrane Molecular Sensing Platform

Wei-Ning Liu, Sunny Shah, Satyajyoti Senapati and Hsueh-Chia Chang, Chemical & Biomolecular Engineering, University of Notre Dame, Notre Dame, IN

We report a new molecular (nucleic acid and protein) sensor platform that is economical to mass produce, label-free, multi-target (>5), rapid (<20 min), selective (SNP) and sensitive (pM)----and is hence most suitable for portable diagnostics. In addition to the difficulty in labeling targets, optical detection also suffers from the small number of simultaneous detectable targets, as most fluorescent reporters occupy a bandwidth of more than 50 nm in the narrow 300 nm visible spectrum. Classical electrochemical sensors, on the other hand, suffer from many robustness issues due to spurious electron transfer reactions and capacitance and conductance signals due to conformation change in the target, probe, or Self-Assembled Monolayers in samples of different ionic strengths and counterion valencies. For both optical and electrochemical sensing, the assay time is often in excess of several hours for < nM concentrations due to the diffusion limitation. Our new membrane sensor alleviates all these difficulties. We exploit the fact that when a DC field is applied across a nanoporous ion-selective membrane, a molecularly small charge polarized layer develops on the depletion side after concentration polarization. The presence of molecules in this polarized layer can hence sensitively change the limiting and overlimiting currents of the membrane. The overlimiting current shift is particularly large (~1 V) because large molecules can change the hydrodynamic slip length and hence suppress the surface vortices responsible for the overlimiting current. Furthermore, oppositely charged small molecule targets can produce a bipolar membrane, like a P-N semiconductor junction, that can produce a large water-splitting signal when a high field develops at the carrier-depleted junction with a reverse bias. By forgoing the method of electron transfer to detect a target, we eliminate the need for SAM and the robustness problems that arise from spurious electrochemical reactions. This new molecular sensing technology also accelerates the assay time because the high field in the polarized layer can attract large nucleic acids by dielectrophoresis and the depletion region can concentrate the charged molecules near the sensor surface in a flowing stream. Our platform utilizes DNA probes or aptamers to achieve high specificity and enable sensitive detection of the target molecule down to pM concentrations. We also used flow to remove non-specifically bound molecules and have developed calibration curves for the proper shear rate for a particular target size and probe length. The entire platform is fabricated onto PDMS or polymer chips, and we have tested it for detection of various targets, including nucleic acid sequences (both DNA and RNA) from 20 bp to >1 kb and several proteins.

(340a) Assessment of the Particle-Particle Influence On the Dielectrophoretic Response of Microparticles

Hector Moncada-Hernandez, Ingenieria Electrica y Computacional, Tecnológico de Monterrey, Campus Monterrey, Monterrey, Mexico; Chemical Engineering, Michigan Technological University, Houghton, MI and Adrienne R. Minerick, Michigan Technological University, Houghton, MI

Assessment of the particle-particle influence on the dielectrophoretic response of microparticles

Hector Moncada-Hernandez[1], Adrienne R. Minerick[2]

[1]BioMEMS Research Chair, Tecnológico de Monterrey, Monterrey, NL, México.

[2]MD-ERL, Michigan Technological University, Houghton, MI, USA.

Advances on microfabrication technologies allow for the development of more complex and specific devices for the characterization and manipulation of particles. Microsystems that probe biological particles for diagnostic, materials, and other applications varying in configuration, complexity, and operating conditions. One electrokinetic tool commonly utilized is dielectrophoresis (DEP) which is based on the polarization of particles in the presence of non-uniform electric fields. This technique has been increasingly employed for the manipulation of bioparticles in miniaturized systems due to the easily observable particle movements with slight differences in dielectric properties using very low potentials over small distances. Further, computational models are employed to simulate the behavior of particles under specific conditions and to optimize the performance of a specific microfluidic device. In order to properly simulate the electrokinetic behavior of biological particles, it is important to determine the particle's dielectric properties, which are most commonly obtained by studying the dielectrophoretic response and electrorotation (ROT) spectra for that specific cell. However, the number of particles in the system or number density is important to consider because particle-particle interactions under a controlled external electric field play an important role on the observed dielectrophoretic response. This is largely overlooked in the dielectrophoresis literature, although the particle-particle interactions affect particle polarizations by changing the local electric field between the particles. This work reports experimentally obtained velocity and negative DEP/positive DEP response of chained particles under a characteristic suspending medium and AC signal. Polarization can be determined as a function of the number of particles interacting together in a pearl-chain formation. In the present work, the velocity of spherical polystyrene microparticles and red blood cells (RBC) due to DEP force are measured in a device with a modified quadruple electrodes configuration. The velocity is compared between a single particle, 2, 3 and 4 particles aligned together in a chain formation along the electric field for both polystyrene beads and cells independently and changing the frequency of the signal in a range of 1 kHz to 80 MHz. Results are tabulated by region in the devices. The results from this work will help to further understand how induced dipole polarizations with different concentrations of particles can affect electrokinetic systems.

(340b) Joule Heating Effects On the Dielectrophoretic Force for iDEP Devices

Roberto Gallo, Cátedra de BioMEMs, Instituto Tecnológico y de Estudios Superiores de Monterrey, Monterrey, Mexico, Rafael Davalos, School of Biomedical Engineering and Sciences, Virginia Polytechnic Institute and State University, Blacksburg, VA and Blanca H. Lapizco-Encinas, PhD, Chemical and Biomedical Engineering, Rochester Institute of Technology, Rochester, NY

Dielectrophoresis (DEP), the movement of particle due to polarization effects in nonuniform electric fields, is a technique that is receiving significant attention for many different applications. Traditionally, DEP has been employed using arrays of microelectrodes, which can have some challenges when handling bioparticles. An alternative is the use of arrays on insulators, instead of electrodes to generate nonuniform electric field. In insulator-based dielectrophoresis (iDEP) an array of insulating structures is straddled by two remote electrodes, the presence of the insulators distorts the distribution of the local electric field, creating zones of higher and lower field intensity. Novel designs that apply this principle are being reported continuously since there are still wide areas of opportunities when referring to selectivity and efficiency. Some of the challenges with iDEP devices is that high potentials are required to achieve effective particle manipulation, leading to the generation of Joule heating in the system. Such phenomenon often produces important temperature increase that directly affects the conductivity of the suspending medium. As a result the local electric field will also be affected. In this work, the temperature effects due to Joule heating by the application of a DC electric potential will be investigated for a channel for iDEP with circular insulating posts. The conductivity, the local electric field, and gradient of the squared electric field important for the magnitude of the dielectrophoretic force were computationally simulated. The temperature increase through the channel decreases the magnitude of the dielectrophoretic force and therefore makes it more difficult to overcome electrokinetics and have a trapping effect on particles.

(340c) Electro-Hydrodynamic Concentration of Neutrally Buoyant Particles in Microliter Droplets

Ezinwa Elele, Yueyang Shen and Boris Khusid, Chemical, Biological & Pharmaceutical Engineering, New Jersey Institute of Technology, Newark, NJ

Microdevices that are based upon manipulation of tiny fluid droplets carrying a biological or chemical content have recently emerged as a promising technology for a wide range of applications. Such devices also referred to as "digital microfluidics" do not require a network of microfabricated channels, valves, mixers, and pumps for moving fluids around on the chip. The open and scalable architecture of digital microfluidics plus readily available techniques for basic droplet operations (dispensing, transporting, splitting, merging) offer the possibility of fabricating simple, easily reconfigurable and low cost devices for large-scale parallel operation of liquid samples. The current focus of the droplet-based technology is on the development of methods to perform generic laboratory procedures inside a tiny droplet. We present a new method for the AC electric field driven concentration and removal of particles from droplets of aqueous polymer solutions. Experiments conducted on suspensions of blue polyethylene microspheres dispersed in a saliva simulant and suspensions of hollow glass particles dispersed in low volatile polyethylene glycol (PEG 200) demonstrated that the proposed electro-hydrodynamic method can operate over a broad range of electrical conductivity and viscosity of suspending fluids.

The work is supported by NASA grant NNX09AK06G

(340e) Electroporation and Material Delivery Mechanism of Nanochannel Electroporation

Wei-Ching Liao, Mechanical and Aerospace Engineering, The Ohio State University, Columbus, OH and L. James Lee, Chemical and Biomolecular, The Ohio State University, Columbus, OH

Cell electroporation is one of the well established non-viral approaches for gene therapy and has been widely used by biomedical researchers. Its transfection efficiency and uniformity have been improved in recent years using microfluidics based single cell electroporation. A new nanochannel electroporation (NEP) was developed recently in our laboratory to realize dosage-controlled delivery that could not be achieved by bulk electroporation (BEP) and microchannel electroporation (MEP). Compared to BEP and MEP, NEP has convection dominated delivery, minimal cell damage, and the capability of delivering large transfection agent. In this study, we investigate the poration and material delivery in the processes of NEP. In the experiment, electroporation mechanisms of NEP were observed through the delivery of quantum dots. Detailed electric field, membrane poration, and material delivery under the pulsed field are calculated by computer simulation. From the calculated, electrophoretic and diffusive transports, transmembrane potential distributions, and poration profiles, NEP is proved to be a unique single cell electroporation tool for does-controlled transfection.

(340f) Investigation of the Thermal and Electrical Impact of Electric Fields On Mammalian Cells Manipulated Using Contactless Dielectrophoresis

Michael B. Sano, School of Biomedical Engineering & Sciences, Virginia Polytechnic Institute and State University, Blacksburg, VA and Rafael V. Davalos, Biomedical Engineering and Sciences, Virginia Tech, Blacksburg, VA

Dielectrophoresis (DEP), the motion of a particle in a non-uniform electric field, has become a robust method for analyzing nano-particles, cells, viruses, and DNA based on their physical and electrical properties. A new technique, contactless Dielectrophoresis (cDEP), separates cells from contact with the electrodes by using fluid electrodes which are isolated from the sample channel by thin insulating membranes. This technique has been used to isolate live mammalian cells from dead cells and beads of the same size as well as discriminate between cells with different metastatic behaviors. A significant effort has been made to expand the originally limited frequency spectrum over which these devices are effective. Current designs are capable of manipulating cells at frequencies between 1 kHz and 1 MHz in low conductivity, physiologically suitable buffers. This wide range of operation allows investigators to manipulate mammalian cells at frequencies close to their first Clausius-Mossotti factor crossover point where the difference between similar cells is the most pronounced. This has enabled the use of cDEP devices for a number of biological applications which analyze and culture of the cells off chip, post sorting. To ensure that secondary assays on these cells are accurate, it is important to minimize the physiological stress experienced by the cells as they are sorted. In this study a combination of finite element modeling and experimental examination is used to investigate the effects that continuous exposure to medium and high frequency electric fields have on mammalian cells.

Multi-physics models of two cDEP devices were constructed in COMSOL which incorporated Joule heating, fluid dynamics, and heat transfer in fluids and solids. The first device contained a continuous sorting geometry with saw tooth features in which cells experience a vertically displacing DEP force. The electric field experienced by the cells cycled in magnitude between high and low as they traveled through regions with and without constrictions, respectively. The second device contained a batch sorting geometry in which cells are attracted to and trapped on insulating pillars within the sample channel. In this geometry, cells experience a constant electric field for the duration in which they are trapped. The time dependent electric field dosages were computed using previously published experimental values. A third three-dimensional model of a single cell suspended in solution was then constructed and these experimental field dosages were applied. The resulting increase in transmembrane potential was then calculated and compared literature values for cellular response and damage.

Master stamps of the two theoretical devices were created on separate silicon wafers using deep reactive ion etching. Polymer replicates were then cast in PDMS with a 10:1 ratio of polymer to curing agent. The polymer replicates were bonded to glass slides after exposure to air plasma for two minutes and stored under vacuum. Prior to experimentation, the fluid electrode channels were filled with phosphate buffered saline with a conductivity of 1.4 S/m. MDA-MB-231 breast cancer cells were suspended in a sucrose solution with a conductivity of 0.1 S/m and driven through the devices using a syringe pump. A high voltage amplifier and series transformer were used to apply voltages with amplitudes up to 300 V_{RMS} between 10 and 800 kHz.

Uptake of membrane impermeable dye was noticed in some experimental conditions indicating that some degree of reversible electroporation was occurring. The degree of dye uptake increased as the applied voltage was increased or as the frequency was lowered. Additionally, a small set of experimental conditions resulted in permanent destabilization of the cell membrane due to irreversible electroporation. These results are consistent with those determined numerically. Current work is focused on evaluation of the expression of heat shock proteins after cells are driven through the devices. However, theoretical calculations predict a minimal increase in the fluid temperature, and expression of these proteins is expected to be negligible.

In this work, we showed that for the majority of experimental conditions, prolonged exposure to alternating electric fields in a cDEP device results in minimal impact on analyzed cells. However, there exists a small subset of parameters which induce temporary changes in the permeability of the cell membrane which must be accounted for when analyzing secondary assays on sorted cells.

(340g) Enhancing Electroporation with Targeted Gold Nanoparticles

Shuyan Huang, Vivek Verma, Yingbo Zu and Shengnian Wang, Chemical Engineering/Institute for Micromanufacturing, Louisiana Tech University, Ruston, LA

The emerging physical gene delivery methods, electroporation in particular, offer surgery-like treatment, quick delivery response, and almost no restrictions on cell type and exogenous material properties. They have successfully been applied to facilitate delivery of various molecular probes and understand biological functions and transport at the cellular level. Conventional electroporation has been reasonably successful while carrying several major drawbacks, including including high electric voltage, large DNA consumption, low transfection efficiency and/or cell viability. Here we use gold nanoparticles (AuNPs) with or without or targeting molecules to improve the performance of electroporation on gene delivery to mammalian cells. AuNPs are used to facilitate the polarization and permeability of the cell membrane when imposing electric pulses. Targeting molecules such as transferin help bring AuNPs and/or DNA close to cells. In this way, the uptake of DNA molecules is enhanced by allowing more molecules to enter the cell cytoplasm before the disappearance of the transient pores and the cell recovery. AuNPs of various sizes, geometries, and concentration are tested. The delivery efficiency was evaluated with both anchor cells (i.e., NIH 3T3) and suspension cells (i.e., K562). Gold nanoparticles show excellent biocompatibility and are facile for surface functionalization. The transfection efficiency was found much enhanced with targeted gold nanoparticles as compared to conventional electroporation. This study provides a new approach on improving the deliver efficiency of nucleic acid or anticancer drugs through the combination of targeted nanoparticles and electroporation.

(394a) Combination of Electroosmotic Flow and Dielectrophoretic Trapping of Particles in a Contactless-Dielectrophoretic Device

Michael B. Sano, School of Biomedical Engineering & Sciences, Virginia Polytechnic Institute and State University, Blacksburg, VA, **Roberto Gallo**, Cátedra de BioMEMs, Instituto Tecnológico y de Estudios Superiores de Monterrey, Monterrey, Mexico, Blanca H. Lapizco-Encinas, PhD, Chemical and Biomedical Engineering, Rochester Institute of Technology, Rochester, NY and Rafael V. Davalos, Biomedical Engineering and Sciences, Virginia Tech, Blacksburg, VA

Microfluidics is a rapidly growing field that offers great potential for many biological and analytical applications. There are important advantages that miniaturization has to offer, such as portability, shorter response times, higher resolution and sensitivity. There is growing interest on the development of microscale techniques. Among these, electrokinetic phenomena have gained significant importance due to their flexibility for handling bioparticles. Dielectrophoresis (DEP), the manipulation of particles in non-uniform electric fields due to polarization effects, has become one of leading electrokinetic techniques. DEP has been successfully employed to manipulate proteins, DNA and a wide array of cells, form bacteria to cancer. Contactless-DEP (cDEP) is a novel dielectrophoretic mode with attractive characteristics. In cDEP nonuniform electric fields are created by using insulating structures and external electrodes that are separated from the sample by a thin insulating barrier, this prevents bioparticle damage, and makes cDEP a technique of choice for many biomedical applications. In this study, a combination of cDEP generated with AC potentials and electrokinetic liquid pumping generated with DC potentials is employed to achieve highly controlled and selective particle trapping and manipulation, allowing for lower applied potential than those used in traditional insulator-based DEP, and making for a simpler system that does not require the use of an external pump. This is the first demonstration of electrokinetic (EK) pumping in which the driving electrodes are not in direct contact with the sample fluid. Multi-physics simulations were used to aid with the design of the system and predict regions of particle trapping. Results show the advantages of combining AC-cDEP with DC EK liquid pumping for dynamic microparticle trapping, release and enrichment.

(394b) ABO Membrane Antigens Alter Dielectric Properties of Red Blood Cells

Kaela M. Leonard and **Adrienne Minerick**, Department of Chemical Engineering, Michigan Technological University, Houghton, MI

Human blood is of incredible diagnostic interest because it is easily obtainable and can provide a plethora of information about a person's general health such as protein, iron, vitamin and electrolyte levels. The goal of this research is to expand blood diagnostic capabilities by using dielectrophoresis as a means to provide an ABO-Rh blood type classification from blood. Prior results from our lab suggest that the dielectric properties (permittivity and conductivity) of human erythrocytes change dependent upon ABO-Rh antigen expression. The Clausius-Mossotti factor is directly dependent upon the dielectric properties and thus alternating current dielectrophoresis is a mechanism by which ABO-Rh blood types can be identified by quantifying red blood cell behaviors. Additional prior results with β(1-3) galactosidase treatment demonstrated cleavage of the ABO antigens without harming the erythrocyte and verification of the galactose residues in the supernatant. Treatment yields immunologically bare erythrocytes whose membrane surface is bereft of the ABO antigens. Previous experiments done in a 0.9S/m conductivity buffer have shown that the range over which the crossover frequency occurs for ABO-Rh blood types in their native state is 17MHz, whereas the range over which the crossover occurs for the modified samples is 5MHz. This uniformity in cross over frequency is an indication that the ABO antigens influence crossover frequency of the red blood cells. Further experiments at lower medium conductivities of 0.01S/m and 0.1S/m have been conducted over a large frequency range In addition to measuring the crossover frequencies of blood samples, the dielectrophoretic spectra as a function of frequency have been constructed using a novel intensity profile approach. This approach allowed typical dielectrophoretic traces to be plotted versus frequency for each ABO-Rh blood type, both native and modified samples. In this manner, not only could the crossover frequencies of the blood types be compared, but also the curvature of their traces, giving a more accurate method of distinguishing between blood types.

(394c) Electrostatics Potential in Annular Geometry

Abbas Motamedilamouki[1], Parvin Golbayani[1] and Dr. Pedro E. Arce[2], (1)Chemical Engineering, Tennessee Technological University, Cookeville, TN, (2)Chemical Engineering, TENNESSEE TECHNOLOGICAL UNIVERSITY, Cookeville, TN

Flow through an annular geometry has many applications in chemical, environmental, mechanical and bio-mechanical engineering [1]. A number of researchers have proposed combining electroosmotic flow (EOF) and pressure-driven flow as a means of controlling the motion and separation of bioparticles in an annular geometry to sort particles using electrophoresis [2-3]. We present here a systematic investigation of the electrostatic potential distribution in an annular geometry. Our objective in this contribution is present a mathematical model for the electrostatic potential distribution in both straight and divergent annular geometry. The analytical solutions for the electric potential profile in the annulus are obtained by solving the 2D Poisson–Boltzmann equation with both long channel and Debye– Huckel approximations. This result is in preparation to the derivation of the Electrohydrodynamic Velocity profile. The ultimate goal of this research is to understand the role of capillary geometry in determining biomolecular separations. As a result of this investigation, one can assess the behavior of the electrostatic potential inside of annular channel. Three key parameters have been identified to describe the electrostatic potential behavior: the angle ( ), ratio of up wall potential to the linear combination of both wall potentials, R, which handles the symmetrical/non-symmetrical aspects of the electrostatic potential, and the ratio of the width to the length (γ) that controls the "shape" of the channel section. Results of this study are illustrated by using a series of portraits that capture the key behaviors of the electrostatic potential with respect to the three parameters described above.

1. Makinde, O. D., Sibanda, P., 2000 Steady Flow in a Diverging Symmetrical Channel: Numerical Study of Bifurcation by Analytic Continuation. *Quaestiones Mathematicae*. 23, 45–57.
2. Xuan X., Xu B., Li D., 2005 Accelerated particle electrophoretic motion and separation in converging–diverging microchannels. *Anal Chem*. 77, 4323–4328.
3. Ai Y., Joo S. W., Jiang Y., Xuan X., Qian S., 2009 Transient electrophoretic motion of a charged particle through a converging–diverging microchannel: effect of direct current-dielectrophoretic force. *Electrophoresis*. 30, 2499–2506.

(394d) Micron-Scale Ion Concentration Gradients in Nonuniform AC Electric Fields

Ran An[1] and Adrienne R. Minerick[1,2], (1)Chemical Engineering, Michigan Technological University, Houghton, MI, (2)Michigan Technological University, Houghton, MI

Micron-scale Ion Concentration Gradients in Nonuniform AC Electric Fields

RAN AN, ADRIENNE R. MINERICK,

Department of Chemical Engineering

Michigan Technological University, Houghton, MI 49931

Erythrocytes were observed to first swell slightly and then shrink substantially over time in phosphate buffer saline in a two dimensional nonuniform AC dielectrophoretic field. We hypothesized that this cell behavior was due to changing osmotic pressure between the cell and the local medium as propagation of an ion concentration gradient occurs in the electric field area. Cells are known to respond rapidly to the tonicity of the surrounding media; isotonic media yields an erythrocyte typical biconcave shape while hypotonic will cause erythrocytes to swell and hypertonic cause cells to shrink. Experimental results were examined to describe the cell shrinkage behavior in initially isotonic buffer solution under nonuniform AC DEP field. Results have shown that cell shrinkage can be observed from 180 seconds close to the higher field density region and will expand to the lower field density region under high frequency (1 MHz) AC field. Also, cell shrinkage starting time will change with the change of applied signal frequency and peak-to-peak potential. Ion behaviors were examined with COMSOL 'Transport of Dilute Species' physics to simulate diffusive, convective and electromigration transport under experimental conditions. Results have shown that high ion concentration increased in the regions adjacent to the electrode over 100 cycles of the AC field. These results have important implications in the field. Research in this field has, until now, assumed polarization of dielectrics in dielectrictrophoretic fields occurs in relatively constant medium conditions. The Debye layer comprises tens to hundreds of nanometers from the dielectric particle surface or electrode particle surface; ion behaviors in this layer have been studied. This work elucidate the formation and transportation of waves of ions tens to hundred of microns from the electrode surfaces in non-uniform AC electric fields and corresponding experimental results suggest complex interaction between the dynamic ion distribution in the medium and the particle's induced dipole response.

(394e) Multi-Frequency Impedance Method for the Rapid Detection of Viable Micro-Organisms

Sachidevi Puttaswamy, Byung-Doo Lee and Shramik Sengupta, Biological Engineering, University of Missouri, Columbia, MO

We present a novel culture-based electrical method for detecting presence of viable micro-organisms in suspensions that is significantly faster than existing methods. In our previous work, we have used our method to detect presence of viable bacteria in various samples, and in this paper, we extend our work to detect viable yeast and mold.

The principle underlying our method of detection is the polarizability of viable microbial cells. Cell membranes of viable microbial cells do not, in general, allow charge-carrying ions to pass through. So, in the presence of an alternating electric field, there occurs a build-up of charge at the membrane, causing the cells to act like capacitors. As they multiply, there occurs a corresponding increase in the charges thus stored in the interior of the suspension (its "bulk capacitance"), and this increase in bulk capacitance serves as our "signature" indicating the presence of live micro- organisms. The bulk capacitance is estimated by first confining an aliquot drawn from the suspension of interest into a long microfluidic channel with electrodes at either end, measuring the electrical Impedance (*Z*) measurements at multiple frequencies (ω) ranging from 1 KHz to 100 MHz, and then analyzing the *Z*vs ω data to obtain an estimate of the bulk capacitance. Both the microfluidic geometry and the high frequencies used serve to overcome the effects of interfacial (double-layer) capacitances at the electrodes, which usually make it difficult to measure bulk capacitance.

The results obtained are used to demonstrate the ability of our method to extend it to detect viable yeasts and molds. We show that, for slow growing yeast and mold (doubling time of 6-8 hrs), our method is able to detect ~5, -10, of *Dekkera anomala and Aspergillus brasiliensis* in growth media in 60 and 45 hrs, respectively, with a threshold concentration of detection of ~5000-8000 CFU/ml.

(394f) Microfluidic Characterization of the Dielectric Properties of Human Mesenchymal Stem Cells, Adipocytes, and Osteoblasts

Tayloria Adams, Chemical Engineering, Michigan Technological University, Houghton, MI and Adrienne Minerick, Department of Chemical Engineering, Michigan Technological University, Houghton, MI

Degenerative diseases such as diabetes and Parkinson's disease combined affect over 28 million people. Human mesenchymal stem cells (hMSCs) can be used to autologously treat such diseases through stem cell purification. hMSCs are ideal because they are easily isolated from bone marrow and they have the ability to self-renew while undergoing differentiation. hMSCs have the potential to differentiate into osteoblasts, adipocytes, chrondrocytes, astrocytes, and myoblasts based on environmental promoters. One issue associated with the differentiation of hMSCs is cell purification. In order for hMSCs and their differentiated progeny to be useful in cell therapies, the cells need to be separated after differentiation occurs. Currently, magnetic-activated cell sorting (MACS) and fluorescence-activated cell sorting (FACS) are the separation techniques employed. FACS and MACS use unique cell-surface antigens or other recognition elements to tag the target cells. This 'labeling' of the cells alters cellular function which is not desirable for the treatment of degenerative diseases. Another disadvantage of FACS and MACS is that it takes days or event weeks for cells to be purified. These approaches require expensive raw materials, are labor intensive, and have seen limited adaptation to other stem cells. Therefore a label-free, one-step cell purification technique that can rapidly sort all types of stem cells without altering cellular function is needed.

Dielectrophoresis (DEP) is a separation technique that has the potential to overcome the short-comings of FACS and MACS. DEP utilizes nonuniform electric fields to polarize cells based on the polarizability and dielectric properties of their membrane, cytosol, and other structurally dominant organelles. Based on these properties, cells will exhibit either positive DEP force, cells move to areas of stronger electric field strength, or negative DEP force; cells are repelled from areas of stronger electric field strength. DEP has been used to study other cell systems such as red blood cells, breast cancer cells, leukemia cells, cervical cancer cells, white blood cells, and yeasts cells. Based on evidence from these cell studies, DEP can be used to determine the dielectric properties of subtle cellular changes, and we hypothesize that DEP can discern hMSCs and morphological changes within its differentiated progeny.

In this work, a microfluidic device with gold quadrupole electrodes spaced 50 microns apart within a microfluidic channel and chamber device is used to quantify the DEP response of hMSCs that were differentiated into osteoblasts and adipocytes. The hMSCs were characterized from 100 kHz to 80 MHz frequency at a rate of 0.67 MHz/sec for 120 seconds in dextrose media at 0.01 S/m to 0.90 S/m conductivities. Narrower frequency sweeps were used to map out ranges with positive, negative, or cross-over frequency DEP behavior of the hMSCs. The positive DEP response of adipocytes and bone marrow derived hMSCs were found in a range of 70-80 MHz and 30-80 MHz respectively. COMSOL simulations and MATLAB were used to fit the data to the multishelled spherical and multishelled ellipsoidal DEP polarization models in order to calculate the structural polarizability and conductivities of hMSCs, adipocytes, and osteoblasts. These preliminary results suggest that DEP is a sufficient method to distinguish different cell types. Electrokinetically identifying differences in stem cells have broad implications in purification and control for tissue engineering and cell therapies for degenerative diseases.

(394g) Supporting Electrolyte Gradients in a 1 Mm DC Electric Field Microchannel

Aytug Gencoglu, Department of Chemical Engineering, Michigan Technological University, Houghton, MI and Adrienne R. Minerick, Michigan Technological University, Houghton, MI

When an electric field is applied across an electrolyte solution, solute anions and cations electrophoretically migrate to the anode and cathode, respectively. As a concentration gradient starts to form, diffusional ion transport begins to counter electrophoretic motion, although this is potentially negligible since diffusional ion mobilities are an order of magnitude or more smaller than the electrophoretic mobilities achieved in microfluidic devices. The electric field catalyzes reactions at the electrodes in an aqueous medium; H^+ and OH^- ions are generated at anode and cathode surfaces, respectively. The association/dissociation of solute species is another mechanism that affects ionic concentrations at a given locality. This especially becomes important when the abovementioned electrode reactions cause a pH gradient. In that case, the extent of buffer species dissociation varies spatially, in accordance with the pH gradient. Thus, the pK values of the solute species is a factor in determining the ion concentration gradients in microfluidic devices under an electric field. Microdevices typically contain <10 nL volumes such that movements of only 1 nanomole of ions can cause significant shifts in local concentrations. Recent results revealed that the H^+ concentrations in a phosphate buffer can go from pH 8 to pH 3 in under 10 mins in a 100 V/cm electric field across a 1 cm microfluidic channel designed for protein iDEP, and under 1 min in a 3000 V/cm field in the same system. Such concentration changes can have a profound effect; for instance, it was discovered that the reported pH changes enabled protein iDEP by increasing particle size through protein aggregation.

The study presented here aims to complement the abovementioned work by elucidating the K^+ and Cl^- concentration behaviors in similar electrokinetically driven microdevices. K^+ is present as a supporting electrolyte and buffer ion in the phosphate buffer saline while Cl^- is present only as a supporting electrolyte. Therefore, due to the buffer association/dissociation mechanism mentioned above, it is expected that the changes in K^+ and Cl^- concentrations do not mirror each other. K^+ and Cl^- concentrations were quantified with fluorescence microscopy using potassium green and MQAE (n-(ethoxycarbonylmethyl)-6-methoxyquinolinium bromide), respectively. The electric fields were applied across a 1 mm gap between Pt thin film electrodes in a microfluidic channel. Fluorescence intensity profiles across the gap were analyzed at 535 (Potassium green) nm and 460 (MQAE) nm, then correlated to concentration profiles. 0.1 S/m phosphate buffer solutions were spiked with KCl at concentrations ranging from 0 to 100 mM and the ion concentration gradient formation was observed under 100 and 3000 V/cm electric fields. The ion concentration gradient formation behavior provides insight into the transient responses of microfluidic devices, the buffering reactions occurring dynamically within the devices, and underscores the substantial differences between the solutions loaded into devices and the rapidly changing solutions within electric fields.

(394h) Dielectrophoretic Differentiation of Bioparticles in a Sawtooth Microchannel

Paul V. Jones and Mark A. Hayes, Chemistry and Biochemistry, Arizona State University, Tempe, AZ

Rapid bioparticle characterization, as a research goal, has fueled a significant and wide-ranging body of work. Much has been accomplished, but new techniques and applications continue to emerge. Innovations in this field may soon enable the development of rapid, on-site bioanalysis devices that improve the availability, accuracy, and scope of clinical diagnosis. Microfluidic electrokinetic approaches provide unique advantages, including short analysis times, microliter sample and reagent volumes, potentially low cost, and portability. The work presented here explores the use of a single, continuous microchannel in which opposing electrokinetic and dielectrophoretic forces create multiple and distinct bioparticle traps. Specifically, it focuses on capture of bioparticles using DC insulator-based dielectrophoresis in a converging, sawtooth-patterned microchannel. The channel design enables localized isolation and concentration of specific particles based on differences in their physical properties. Various targets with disparate characteristics have been captured and concentrated within the device, including human blood cells and mature amyloid protein fibrils. In each case, reproducible capture occurred at specific locales within the sawtooth channel. Differentiable capture of three live *Escherichia coli* serotypes has also been demonstrated, indicating the potential this technique holds for diagnostic applications.

(394i) Investigating the Electrical Properties of Prostate Cancer Cell Lines Using Contactless Dielectrophoresis

Alireza Salmanzadeh[1,2], Mohammad Bonakdar[1,3], Michael B. Sano[1], Lina Romero[4], Rafael V. Davalos[1] and Scott D. Cramer[4], (1)Biomedical Engineering and Sciences, Virginia Tech, Blacksburg, VA, (2)Engineering Science and Mechanics, Virginia Tech, Blacksburg, VA, (3)Mechanical Engineering, Virginia Tech, Blacksburg, VA, (4)Department of Pharmacology, University of Colorado Denver, Aurora, CO

Prostate cancer is the most common malignancy in men and the second cause of cancer death among men in the United States. An early diagnosis of this cancer is crucial to decrease the mortality rate due to this cancer. Investigating the electrical and mechanical properties of cancer cells may lead to finding novel techniques to diagnose cancer in early stages and new cancer treatment methods. Dielectrophoresis (DEP), the polarity induced motion of particles in a non-uniform electric field, is a successful technique for characterization and separation of cells. The non-uniform electric field is traditionally generated by patterning metal electrodes onto the bottom of a microfluidic channel. While this technique has been tremendously successful, electrode delamination and fouling are a challenge. We have developed a relatively new technique, known as contactless dielectrophoresis (cDEP), in which metal electrodes are exchanged for conductive fluid electrode channels. These fluid electrode channels are isolated from a main sample channel by means of a thin insulating membrane. The absence of direct contact with metal electrodes provides a permissive environment for the cells and allows for studying their electrical properties.

We previously characterized the dielectrophoretic response of prostate tumor initiating cells (TICs) of a prostate cancer cell line (PC3) utilizing cDEP. An electric field generated within the cDEP microdevice using applied voltages between 0 and 300 V_{RMS} with frequencies between 200 and 600 kHz. We have shown that the required voltage to completely trap TICs is different from that of non-TICs. This data was then used to find the optimal parameters to sort then culture enriched TIC populations. We observed that the TIC-identified cells, sorted using cDEP produced significantly more spheroids than control. Additionally, the average size of the cDEP enriched TIC spheroids was about 4 times larger than control, and about 17 times larger than cDEP enriched non-TICs spheroids.

In this study, we investigated the dielectrophoretic properties of three different prostate cancer cell lines which have previously been evaluated in the study of prostate cancer progression: PC3, DU145, and LNCaP. The LNCaP cell line is established from human prostate adenocarcinoma and derived from the lymph node metastasis. The PC3 cell line has a high metastatic potential and was derived from the bone metastasized prostate and the DU145 cell line was derived from the brain metastasized prostate cancer.

A low frequency cDEP device was used to find the crossover frequency of these cell lines. A silicon master stamp was first fabricated using typical photolithography techniques and deep reactive ion etching. The device was then cast in PDMS using a 10:1 ratio of monomers to curing agent. The polymer replicates were bonded to glass slides using air plasma after access holes were punched. The cells from three lines were independently suspended in a low conductivity buffer and driven through the device using a syringe pump at a flow rate of 0.005 mL/hour. A 200 VRMS signal was then applied across the device for frequencies between 5 and 100 kHz. The distribution of cells within the device was recorded and processed to determine the first crossover frequency.

We found that the crossover frequency for PC3, DU145, and LNCaP are 23.23 ± 0.51, 22.68 ± 0.64, and 31.22 ± 1.30 kHz, respectively. Crossover frequencies of PC3 and DU145 were not statistically different but both of them were significantly different from LNCaP ($p<0.001$ and $p<0.005$, respectively). We hypothesize that being derived from different types of metastasis could be the main reason for difference in their DEP response.

(394j) Dielectrophoretic Response of Polystyrene Particles and Perfluorocarbon Oil-Core, Chitosan/Poly-L-Lysine/CaPO4 Shell Nanoparticles

Chungja Yang, Chemical Engineering, Michigan Technological University, Houghton, MI, Chun-Jen Wu, Materials Science and Engineering, University of Utah, Salt Lake, UT, Agnes E. Ostafin, Materials Science & Engineering, University of Utah, Salt Lake City, UT and Adrienne R. Minerick, Michigan Technological University, Houghton, MI

Dielectrophoretic Response of Polystyrene Particles and Perfluorocarbon oil-core, Chitosan/Poly-L-lysine/$CaPO_4$ shell Nanoparticles

Chungja Yang[1], Chun-Jen Wu[2], Agnes E. Ostafin[2], and Adrienne Minerick[1]

[1]*Chemical Engineering Department, MTU*

[2]*Materials Science and Engineering Department, University of Utah*

Dynamic (dis)assembly of nanoparticles into three-dimensional, packed structures would be useful for drug delivery, nanoparticle films and diagnostics. Dielectrophoretic (DEP) microdevices can rapidly assemble, discriminate, and manipulate particles based on a polarizability and medium interactions within non-uniform electric fields. DEP has been used to discern micrometer scale cells and particles, but the manipulation of nanometer scale objects such as viruses, proteins, and nanoparticles has seen less exploration due to the lower forces on these smaller particles. In this work, we examine dielectrophoretic behaviors of spherical core-shell nanoparticles (CSnp) in 2D and 3D particle-assemblies in order to determine dielectric properties (permittivity and conductivity) of the core and shell. Three types of CSnp were custom synthesized with a perfluorocarbon oil liquid core within a phospholipid micelle templating for three shell materials (chitosan, poly-L-lysine, and $CaPO_4$). The synthesis procedures for all three biocompatible shell materials are established and demonstrate a range of permittivities (chitosan= $7.1*10^{-10}$-$4.4*10^{-9}$ F/m, poly-L-lysine= $6.2*10^{-10}$-$8.0*10^{-10}$ F/m, $CaPO_4$= Unknown) and conductivities (chitosan= $3.0*10^{-4}$- $3.5*10^{-1}$ S/m, poly-L-lysine, $CaPO_4$= Unknown) based on concentrations and frequencies. We report the frequency-dependent and medium conductivity-dependent responses of ~250 nm CSnps with ~10 nm shells for all three shell types. Experiments were conducted within a 100 nl chamber housing 100 um wide Ti/Au quadrupole electrodes spaced 25 um apart. Frequencies from 100kHz to 80MHz at a fixed local field of $10V_{pp}$ were tested and the frequency-dependent DEP responses of the nanoparticles were quantified by tracking optical intensity profiles via video microscopy. Average velocity of particles inferred from intensity profiles moving up the electric field gradient (positive DEP) and down the electric field gradient (negative DEP) were compiled as a function of frequency and medium conductivity, then compared with spherical core-shell models for DEP polarization in order to calculate core and shell dielectric permittivity and conductivity for comparison with the literature. This data is useful for dynamic assembly applications for these biocompatible nanoparticles.

(394k) A Combinatorial Microfluidic Approach for Point-of-Care Applications

Ritika Mohan[1], Arnab Mukherjee[2], Charles M. Schroeder[2] and Paul J.A. Kenis[3], (1)University of Illinois, Urbana Champaign, Urbana, IL, (2)Chemical and Biomolecular Engineering, University of Illinois at Urbana-Champaign, Urbana, IL, (3)Department of Chemical & Biomolecular Engineering, University of Illinois at Urbana Champaign, Urbana, IL

We describe a microfluidic platform with optimized normally closed valves for high throughput quantitative analyses of stimulus driven responses in *E. coli* cells [1]. Specifically, we designed and fabricated a 4x6-array microfluidic chip capable of confining cells and reagents in sub-micron volume compartments. Leak-free isolation of contiguous compartments is achieved with elastomeric valves that are closed at rest [2]. Actuation of valves with negative pressure permits controlled exposure of cells confined in ~2 nL volume compartments to various chemical stimuli loaded in adjacent compartments. The device design is convenient for biological assays due to increased detection sensitivity, thereby enabling analysis at single cell resolution. Additionally, the device facilitates portability and exhibits multiple advantages as a miniaturized biological assay, including low sample and reagent volumes for analysis and integration of assay steps in an *on-chip* format. The overall goal of this work is to develop a diagnostic platform for screening antibiotic resistant phenotypes in infected clinical samples.

In this work, we employ the microfluidic platform to investigate antibiotic susceptibilities of *E. coli* cells to several commonly used antibiotics. Furthermore, we explore the synergistic and antagonistic effects of antibiotic combinations on *E. coli* cell proliferation. Specifically, we investigate the effects of ampicillin, tetracycline, chloramphenicol, cephalexin, ciprofloxacin, and combinations thereof on actively growing *E. coli* cells. The results were used to determine effective bactericidal and bacteriostatic concentrations for the aforementioned antibiotics (*e.g.*, > 100 μg/mL for ampicillin, > 10 μg/mL for tetracycline). We also observed and characterized synergistic effects of certain antibiotic combinations. For example, although ampicillin at 10 μg/mL and tetracycline at 1 μg/mL are ineffective individually, together they exhibit significant bactericidal activity when used in combination. Finally, we extended the platform to characterize antibiotic resistance in a model pathogen, *Pseudomonas aeruginosa*.

Overall, these results emphasize the utility of the microfluidic platform for rapidly characterizing antibiotic susceptibilities over a wide range of concentrations. The platform improves on existing diagnostic methods in terms of assay time, ease-of-use, sensitivity, and sample volumes [3]. The described microfluidic platform is expected to expedite elucidation of treatment regimens necessary for combating infections with multi-antibiotic resistant "superbugs" that represent an emerging and acute global health concern.

References:

[1] R Mohan *et al.*, *MicroTAS*, **2011**, Seattle, USA.

[2] R Mohan *et al.*, *Sensors and Actuators B*, **2011**, 1, 1216-1223.

[3] JP Torella *et al.*, *PLoS Comput Biol*, **2010**, 6.

(394l) Assembly of Anisotropic Particles Under Electric Fields

Fuduo Ma, Sijia Wang and Ning Wu, Chemical Engineering, Colorado School of Mines, Golden, CO

Anisotropic colloids can possess anisotropic properties in geometry, chemical composition or surface functionality. As such, they can recognize each other and assemble via directional interactions. By tuning the directional interactions between particles, anisotropic particles can assemble into diversified types of crystalline and non-crystalline structures. This poster will focus on the electric-field directed assembly of two types of anisotropic particles -rigid dimers and flexible trimers under electric fields.

For colloidal dimers with rigid bonds, we recently observed rich patterns in both crystalline and non-crystalline forms, such as clusters with chiraities, and dimer crystals with alternating orientations, which depend on the anisotropies in geometry and surface charges. Our modeling demonstrated that the competition and balance between electrostatic interactions and electrohydrodynamic interactions bring a variety of phases and assembly pathways between dimers. The close-packed crystal of dimers provides promise for fabricating three-dimensional photonic crystals based on non-spherical particles.

In addition, we have investigated colloidal trimers with flexible bonds, assembled from spherical colloids under electric fields. At low particle and salt concentrations, we observed a family of well-defined clusters ranging from 3 to 9 particles, with non-trivial high populations for trimers, tetramers, hexamers, and nonamers. At higher particle and salt concentrations, the colloidal clusters with flexible angles can further assemble and connect themselves into several hierarchical structures, including non-close-packed networks that have not been observed before. The dynamic assembly process of colloids resembles the chemical reactions between real molecules, which could open a new realm for making complex "colloidal molecules" from simple particles with prescribed assembly pathways.

(394m) Solvent Compatible Polymer Based Microfluidics for Applications in Pharmaceutical Industry

Sachit Goyal, Amit V. Desai and Paul J.A. Kenis, Department of Chemical & Biomolecular Engineering, University of Illinois at Urbana Champaign, Urbana, IL

Poly(dimethylsiloxane) or PDMS has been the prime material of choice in microfluidics mainly because PDMS enables simple, inexpensive fabrication using soft-lithographic technique, which facilitates rapid prototyping of devices. Other advantages, such as biocompatibility, optical transparency, low young's modulus, have led to the use of PDMS-based microfluidics for several applications in chemistry and biology. However, PDMS suffers from several limitations, including poor solvent compatibility, small molecules absorption, and limited compatibility with analytical techniques, low mechanical rigidity, and issues with large-scale fabrication. Traditionally, silicon and glass have addressed some of PDMS' deficiencies, but the complex and expensive fabrication procedures, material fragility, and incompatibility with a wide range of analytical techniques have limited their applications.

Polymeric materials that enable simple, inexpensive fabrication while addressing PDMS limitations are emerging as attractive candidates for developing microfluidic devices. Since no single polymer will satisfy all the requirements of the end application, we have developed hybrid devices comprising multiple layers of different polymers to overcome the deficiencies resulting from using single polymers. Our hybrid-microfluidic platforms, ranging from simple single channel devices to complex multilayer devices, comprise thin layers of different polymers, *viz.* cyclic olefin copolymer (COC), Teflon FEP, perfluoropolyether (SIFEL), thiolene, and PDMS.

Here, we demonstrate the application of these devices in the pharmaceutical industry for two specific examples. In the first example, we developed microfluidic platforms for *screening solid forms (salt, cocrystal and polymorph) of pharmaceutical compounds* with different additives (*e.g.*, salt or cocrystal formers) employing several modes of crystallization. These platforms allow for solid form screening much early in the drug development cycle when only small quantities of drug (~10 mg) are available thereby aiding in expediting the drug development process. The platforms consist of multiple layers of COC-PDMS-PDMS-COC-PDMS-thiolene and COC-PDMS-SIFEL-SIFEL-Teflon-SIFEL-thiolene. In the second example, we have developed devices for *liquid-liquid extraction*, where (a) thiolene-glass devices were used to evaluate distribution coefficient of drugs in octanol-water mixture for *in-vitro* drug uptake studies and (b) COC-SIFEL-SIFEL-glass devices were used to purify radioisotopes with applications in synthesis of cancer imaging agents. The material choice was guided by several considerations, including compatibility with solvents (*e.g.*, alcohols, toluene, DMSO, chloroform) and analytical techniques (X-ray, Raman), the strength of bond between heterogeneous polymers bonded via surface activation by plasma or chemical treatment, amenability for high-resolution fabrication, and mechanical robustness as well as reliability of the assembly for long-term experiments.

(394n) Frequency Dependence of Protein Dielectrophoresis Probed with Insulator Based Devices

Fernanda Camacho-Alanis, Asuka Nakano, Tzu-Chiao Chao and Alexandra Ros, Department of Chemistry and Biochemistry, Arizona State University, Tempe, AZ

Dielectrophoresis (DEP) occurs when polarized particles in a non-uniform electric field move towards (positive DEP) or away (negative DEP) from high electric field gradients. When applied to biomolecules, DEP has great potential as a bioanalytical tool for pre-concentration, fractionation, and separation. However, in contrast to well-characterized biological cells, the mechanism of protein DEP is not well understood limiting its potential for bioanalytical applications. The use of insulator posts (i-DEP) is a novel method that generates inhomogenous electric field gradients within a microchannel. In order to enhance the DEP force on proteins, we combine optical lithography with focused ion beam milling (FIBM) to build an array of nanostructured insulator posts in a microfluidic channel and study the DEP behavior of immunoglobulin G (IgG) molecules under DC and AC conditions. Under DC conditions, IgG molecules move according to positive DEP as reported previously [1]; however, utilizing AC, we observed a change from positive to negative DEP. Our preliminary results show that at 1Hz, IgG molecules concentrate in regions of high electric field gradients indicating positive DEP, however above 5Hz, the particles move away of these regions showing negative DEP behavior. This study indicates that such a novel fabrication process has the potential to improve applications for dielectrophoric separation, concentration, and fractionation of biomolecules.

[1] Nakano A.; Chao T.-C.; Camacho-Alanis F., Ros A. Immunoglobulin G and bovine serum albumin streaming dielectrophoresis in a microfluidic device. *Electrophoresis* 2011, 32, 2314-2322.

(420a) Programmed Assembly of Janus, Patchy and Mixed Strongly Interacting Particles by Electric Fields

Sumit Gangwal[1], Bhuvnesh Bharti[2], Gerhard H. Findenegg[2] and **Orlin D. Velev**[1], (1)Department of Chemical and Biomolecular Engineering, North Carolina State University, Raleigh, NC, (2)Department of Chemistry, Technical University Berlin, D 10623 Berlin, Germany

We will discuss how directional polarization interactions induced by external AC fields can be used to assemble complex particles or particle mixtures into unusual lattices of programmed structure and symmetry. Metallodielectric Janus particles could be driven into new types of staggered chains and anisotropic crystals. Patchy metallodielectric particles in AC fields form networks and crystals of unusual symmetry by quadrupolar and hexapolar interactions. The fundamental interpretation of these results is based on calculations of the electrostatic interactions and simulations of the molecular dynamics of the particles. We will also report a new class of permanently bound structures field-assembled from binary mixtures of microspheres of opposite charge with strongly attractive interactions. We demonstrate that the assembly process critically depends on the size ratio and the number ratio of the two types of particles, and derive assembly rules for the structures resulting from the process. The statistical model that we developed for this new type of irreversible directed assembly predicts a chain length distribution that was found to be in excellent agreement with the experimental data. These findings could form a basis for tailored assembly of supracolloidal structures from complex binding units.

(420b) Nanoparticle Directed Assembly Using Electric Fields

Eric M. Furst, Department of Chemical and Biomolecular Engineering, University of Delaware, Newark, DE

Nanoparticles are exceptionally promising components for engineering advanced functional materials and devices – this potential arises from their distinct chemical and physical functionalities (optical, electronic, catalytic, etc.) that stem from their near-atomic nature and collective behavior when particles come together into a coherent spatial organization. Such organization may be an equilibrium or non-equilibrium structure. Engineering nanoparticle assembly across a wide hierarchy of length scales, spaning orders of magnitude, from nanometers to macroscopic dimensions, is the key barrier towards harnessing this transformational technology. Fortunately, the combined use of sophisticated particles that encode self-assembly information through shape or interactions with directing fields, such as fluid flow and electromagnetic fields, has proven to be an effective organizing principle [1,2]. Directed self-assembly has the potential to produce functional structures with novel properties while retaining the essential features of a highly scalable, bottom-up fabrication process. In this talk, I will discuss recent examples of nanoparticle assembly, focusing on the use of directing electric fields to achieve close-packed structures encoded by the constituent nanoparticle shape.

1. Glotzer, S. C. and Solomon, M. J. Nat. Mater. 6, 557–562 (2007).

2. Grzelczak, M., Vermant, J., Furst, E. M. and Liz-Marzán, L. M. ACS Nano 4, 3591–3605 (2010).

(420c) A Hexatic-to-Disorder Transition in Colloidal Crystals near Electrodes: Stronger Flow Yields Less Order

Cari S. Dutcher, Nicholas H. Talken and **William D. Ristenpart**, Chemical Engineering and Materials Science, University of California Davis, Davis, CA

Colloids are known to form two-dimensional, hexagonal closed packed (HCP) crystals near electrodes in response to electrohydrodynamic (EHD) flow. Previous work by several groups has established that the strength of the EHD flow increases as the applied AC frequency decreases, suggesting that the driving force for crystallization should increase at lower frequencies. Here we report that the HCP crystals instead undergo an order-to-disorder transition at sufficiently low frequencies, despite the increase in the attractive EHD driving force. At large frequencies (~500 Hz), monodisperse suspensions of micron-scale particles are observed to arrange into planar HCP crystals, consistent with prior work. As the frequency is decreased below approximately 250 Hz, however, the crystalline structure transitions to randomly close packed (RCP) with an orientational order parameter, ψ_6, significantly less than one. The transition is reversible and second order with respect to frequency, and independent measurements of the EHD aggregation rate confirm that the EHD driving force is indeed higher at the lower frequencies. We present evidence that the order-disorder transition is instead caused by an increased particle diffusivity associated with a corresponding increase in the particle height over the electrode induced at lower frequencies. The observations reported here thus represent a rare instance of an electrically tunable HCP-to-RCP transition, with broad implications for the use of time-varying frequencies to facilitate annealing of planar colloidal crystals.

(420d) New "Chemistry" of Colloidal Particles Induced by the Electric Field - Surprisingly New Results from an Old Experiment

Fuduo Ma and **Ning Wu**, Chemical Engineering, Colorado School of Mines, Golden, CO

Spherical colloids with isotropic properties have been used as building blocks to assemble a variety of 2D and 3D structures in past, such as FCC, HCP, and BCT crystals. We recently, however, have observed surprisingly new types of 2D and 3D structures under the influence of electric field on the surface of a conducting substrate. At low particle concentrations, the spherical colloids can form a series of well-defined colloidal clusters, ranging from 3 to 10. Those clusters can change bond angles freely while maintaining the overall structures intact. The analysis of cluster distributions shows strong peaks for trimer, tetramer, hexamer, and nonamer. In addition, a few types of isomers were observed for the haxamers and their population distributions are non-trivial. At higher particle concentrations, those colloidal clusters with flexible bond angles can further assemble and connect themselves into a good variety of hierarchical structures and non-close-packed networks that have not been observed before. The dynamic process of the assembly resembles closely to chemical reactions between molecular species. We will describe the experimental conditions and formation mechanisms for those surprisingly new structures. The potential applications in photonic crystal, plasmonic clusters, or filtration will also be discussed.

(420e) Direct Numerical Simulations of Colloidal Assembly by Electrophoretic Deposition

Jae Sung Park and **David Saintillan**, Department of Mechanical Science and Engineering, University of Illinois at Urbana-Champaign, Urbana, IL

Motivated by applications in material nanomanufacturing, we perform direct numerical simulations of electrophoretic deposition (EPD) of colloidal suspensions, with the aim of developing a quantitative model of the EPD process and of helping the design and optimization of experiments. A simulation method is developed to model the full deposition process that captures linear electrophoresis, dipolar interactions, electrostatic repulsion and van-der-Waals interactions (based on DLVO theory), Brownian motion, as well as electric and hydrodynamic interactions with the electrodes. Using a fast algorithm, suspensions of up to 5,000 particles are simulated, and results are reported for the final deposit microstructure as a function of effective electric Peclet number, strength of van-der-Waals interactions, Debye layer thickness, and suspension volume fraction. The final microstructure of the deposits is characterized by calculating radial distribution functions, distributions of coordination numbers, and Voronoi diagrams, and results are analyzed in the light of recent experimental observations. We also investigate the use of patterned electrodes and the influence of boundary conditions for the manufacturing of three-dimensional deposits of controlled shape and microstructure.

(420f) Large Scale Electric Field Array Device for Directed Self-Assembly of Multilayer Nanoparticle Materials

Michael J. Heller, Nanoengineering, University of California San Diego, La Jolla, CA and Youngjun Song, Electrical and Computer Engineering, University of California San Diego

AIChE 2012

Topical 3: 2012Annual Meeting of the American Electrophoresis Society

T3009 "Electrokinetic Behavior of Micro/Nanoparticles: Directed Assembly Under Electric Fields"

Large Scale Electric Field Array Device for Directed Self-Assembly of Multilayer Nanoparticle Materials

Michael J. Heller1,2 and Youngjun Song3

University of California San Diego, Department of Nanoengineering1, Department of Bioengineering2 and Department of Electrical and Computer Engineering, Atkinson Hall, Rm 2312, La Jolla, CA 92093-0448

Abstract - In earlier work, an electronic microarray device was used to carry out directed self-assembly of higher order 3D structures from biotin/streptavidin, DNA and enzyme derivatized nanoparticles. Structures with up to fifty layers of alternating biotin and streptavidin, DNA and enzyme nanoparticles were fabricated using a 400 site CMOS microarray system with 50 micron electrode structures. In this process, reconfigurable DC electric fields produced by the microarray were used to rapidly transport, concentrate and accelerate the binding of 40 and/or 200nm nanometer bio-derivatized nanoparticles to selected sites on the microarray. The nanoparticle layering process takes less than one minute per layer. The nanoparticle addressing/binding process was monitored by changes in fluorescence intensity as each nanoparticle layer was deposited. The final multilayered 3-D structures were about two microns in thickness and 50 microns in diameter. In order to make nanoparticle layered materials with macroscopic x-y dimensions (centimeter sizes), a large scale (4") silicon wafer based electronic array device has now been designed and fabricated. The overall goal is to use the large scale device to produce final lift-off photovoltaic/battery prototype materials composed of layered nanoparticles and conductive polymers.

Overview - One of the grand challenges in nanotechnology is the development of fabrication technologies that will lead to cost effective nanomanufacturing processes. In addition to the more classical top-down processes such as photolithography, so-called bottom-up processes are also being developed for carrying out self-assembly of nanostructures into higher order structures, materials and devices. To this end, considerable efforts have been carried out on both passive and active types of Layer-by-Layer (LBL) self-assembly processes as a way to make three dimensional layered structures which can have macroscopic x-y dimensions. Nevertheless, limitations of passive LBL and as well as active assembly processes provide considerable incentive to continue the development of better paradigms for nanofabrication and heterogeneous integration. Electronic arrays have several important features that make them attractive for assisted self-assembly nanofabrication [1]. First, a permeation layer or porous hydrogel is used to cover the microelectrode structures on the array. The permeation layer is usually impregnated with streptavidin which allows biotinylated DNA (antibodies, nanoparticles, etc.) to be bound at the selected site. This layer also allows relatively high DC electric field strengths to be used for rapid electrophoretic transport of molecules and nanostructures, while protecting the more sensitive DNA, proteins (enzymes) or nanostructures from the adverse effects of the electrolysis products generated at the electrodes. A second feature of electronic array devices is that they can be designed in a wide variety of shapes and sizes. Previously, arrays have been fabricated in sizes from 2 mm x 2 mm to over 2.5 cm x 2.5 cm, with 25 to 10,000 electrodes and with electrode structures which range in size from 10 microns to several millimeters. Using a 400 site CMOS microarray device and controller system we have demonstrated rapid and highly parallel assisted self-assembly of biotin

and streptavidin and DNA derivatized nanoparticles into higher order structures [2-5]. In the earlier DNA nanoparticle work, two different sets of streptavidin nanoparticles and the electronic microarray binding sites were derivatized with target and complementary 24mer and 51mer DNA oligonucleotides

(Figure 1A-C). Through a rapid series of DC electric field (electrophoretic) directed depositions, accelerated hybridizations and washing steps, a layered nanostructure material was assembled using complementary DNA oligonucleotides as a structural binding material. The 400 site CMOS microelectrode array device used in the experiments allows individual control of activation time, DC current, and voltage levels and microelectrode polarity (positive, negative, and neutral) (see Figure 1A). The complementary DNA derivatized nanoparticles were specifically concentrated and hybridized (in zwitterionic histidine buffer) to the target DNA sequences bound to the porous hydrogel surface above the microelectrodes. The ability to independently control all 400 sites on the microarray allows for parallel combinatorial testing of a wide range of DNA binding conditions to determine the optimal parameters for hybridization and nanoparticle layering. Previously developed combinatorial methods used to determine optimal conditions for biotin/streptavidin nanoparticle layering [3] were again used to determine the optimal conditions for the DNA nanoparticle layering, i.e., the optimal addressing times and DC current levels [4]. We now show the design and fabrication of a large scale (4") silicon wafer based electronic array device for carrying the directed assembly of multilayered nanoparticle structures.

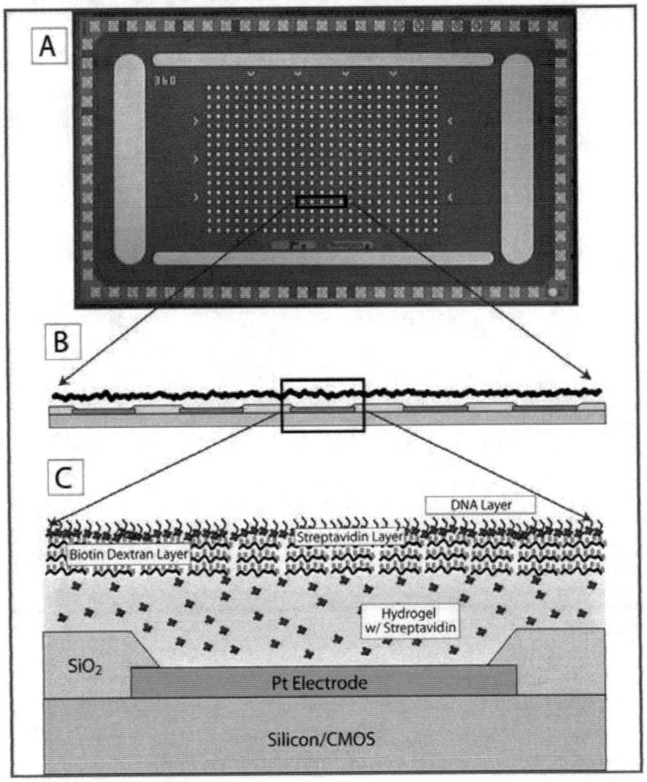

Figure 1 - (A) Shows the 400 site CMOS microelectrode array device. (B) shows a cross section of the 400 site microarray which includes: a silicon base, CMOS control circuitry, three platinum microelectrodes (55 um diameter). (C) shows an expanded view of the cross section with the overlaying polyacrylamide and streptavidin permeation gel layer; a biotin-dextran layer and a final streptavidin layer to which the biotinylated 24mer oligonucleotide target sequences or 51mer oligonucleotide target sequences are attached.

Results: The new 4" silicon wafer electric field directed layer by layer (EFD-LBL) assembler array device is shown below in Figures 2. The total deposition active array area is 5cm2 with 20 by 20 electrodes, the electrode gaps are 8um. The array is covered with a hydrogel (2% agarose gel).

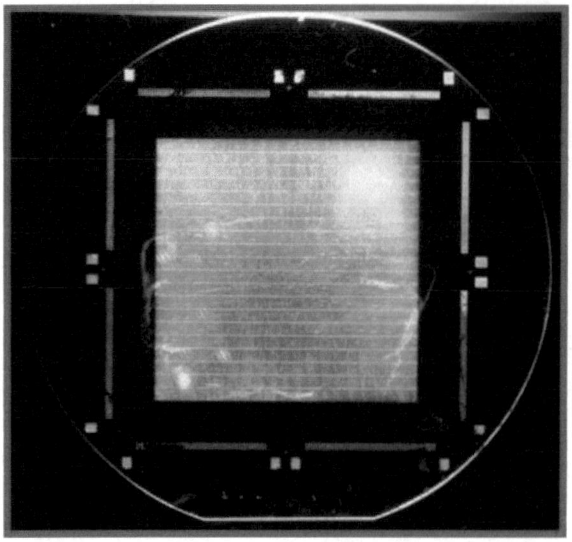

Figure 2 -- The new 4" silicon wafer EFD-LBL assembler array device.

After completing the fabrication of six new 4" large scale EFD-LBL assembler array devices and overcoating them with 2% agarose, micro/nano particle layering experiments were carried out. Figure 3 below show the results for checkerboard deposition of negatively charged 10 micron polystyrene microparticles onto the array device. Each of the 2.5mm X 2.5mm electrodes were biased either (+) positive or (-) negative in a checkerboard fashion. It can be seen in Figure 3 that more of the negatively charged 10 μm polystyrene microparticles have collected where the electric field was positive. Particles on the negatively charged electrodes are non-specifically adhering to the surface.

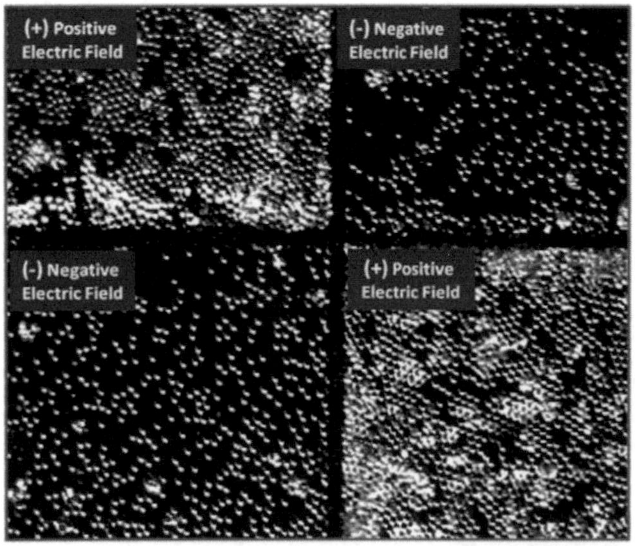

Figure 3 -- Checkerboard deposition of negatively charged 10μm polystyrene microparticles on the array device surface.

In other experiments, positive results were achieved for checkerboard deposition of negatively charged fluorescent 20nm polystyrene nanoparticles. In this case, the nanoparticles could clearly be seen with the epifluorescent microscope concentrating onto the positively charged electrodes. However, when photographed the residual fluorescent from nanoparticles in the solution above the negatively charged electrodes is still visible, even though nanoparticles are actually not on the electrode surface.

CONCLUSIONS

We believe the results of this study carry significant implications for the future use of large scale EFD-LBL devices to carry out the directed assembly of nanoparticles and other derivatized nanocomponents into higher order structures.

EFD-LBL devices and processes using derivatized nanoparticles represents a unique synergy of combining the best aspects of "top-down" and "bottom-up" technologies into a viable nanofabrication process. It also represents a bio-inspired logic for self-assembly and hierarchal scaling of nanocomponents into integrated microscopic structures. In addition to producing lift-off photovoltaic/battery prototype materials composed of layered nanoparticles and conductive polymers, other potential applications for this process include fuel cells, miniaturized chemical and biosensor devices, "micronsize" dispersible chem/bio sensors for environmental and bioagent detection, lab-on a-chip devices and in-vivo diagnostic/drug delivery systems.

REFERENCES - [1] C. Gurtner, E. Tu, N. Jamshidi, R. Haigis, T. Onofrey, C.F. Edman, R. Sosnowski, B. Wallace, M.J. Heller, Electrophoresis 2002, 23, 1543-1550; [2] Dehlinger DA, Sullivan B, Esener S and Heller MJ, "Directed hybridization of DNA nanoparticles into higher order structures", NanoLetters, 8, pp. 4053-4060, 2008; [3] Dehlinger DA, Sullivan B, Esener S, Hodko D, Swanson P and Heller MJ, Automated Combinatorial Process for Nanofabrication of Structures Using Bioderivatized Nanoparticles, J. Assoc. Lab Automation, v.12, #5, pp.267-276, October 2007; [4] Dehlinger DA, Sullivan BD, Esener S and Heller MJ, Electric Field Directed Assembly of Biomolecular Derivatized Nanoparticles into Higher Order Structures, SMALL, V3, #7, pp. 1237-1244, 2007; [5] Hsiao AP and Heller MJ, Electric Field Directed Self-Assembly of Active Enzyme-Nanoparticle Structures, J. Biomedicine and Biotechnology, Volume 2012, Article ID 178487, 9 pages, doi:10.1155/2012/178487

Rapid Electrokinetic Patterning (REP) is a relatively new technique useful for gathering, sorting, and manipulating colloids at an electrode surface. REP requires an AC electric field and a heat source. The AC electric field works to trap particles at a planar electrode surface, and the combination of the AC electric field and the heat source generate a vortex useful for manipulating trapped particles. The heat source used was an infrared laser, which was absorbed by one of the Indium Tin Oxide (ITO) electrode surfaces. The heat absorbed by the surface was transferred to the medium, creating a gradient in the electrical properties of the medium. The AC field acted on this gradient to produce a net force, inducing a vortex. The vortex creates drag and lift forces on the particles, which can be used to sort the particles.

The ability to collect colloids, move them, and control the geometry of the captured group of colloidal particles was explored. One- and two- micron polystyrene particles were used in experimentation. It is shown that simple geometries, such as points and straight lines, can be formed and controlled. The time required to gather particles is explored, as well as the maximum moving speed of the geometries. Groups of particles will follow the laser up to certain speeds, past which the particles are not dragged with the vortex around the focal point of the laser. If speeds are increased far enough past this point, simple geometric shapes form.

The angle of intersecting geometry is explored in order to give insight to the nature of the vortex, as well as two separate geometries brought together. Scanned geometries exist due to the vortex being spread over a larger area due to the movement of the laser. Two separate geometries are shown to exist with the use of one scanning laser. The nature of the aggregations as they interact with each other has been previously unexplored. The angle of intersecting geometry is explored in order to give insight to the nature of the vortex, as well as two separate geometries brought together.

Also explored are closed geometries, such as elliptical shapes. It is shown that one point aggregation can be manipulated into a line-shaped aggregation, and then into an elliptical aggregation. It is further shown that a slip plane can be generated over elliptical geometry, such that a two-dimensional ellipse is patterned to the electrode, while particles in a second layer slip around the top of the patterned ellipse.

(420h) Low-Frequency Dielectric Response of a Single Particle in Aqueous Suspensions

Jingyu Wang and H. Daniel Ou-Yang, Physics, Lehigh University, Bethlehem, PA

For a charged colloidal particle in aqueous suspension, an electric double layer composed of absorbed and attracted counterions, forms at the interface between the particle and the liquid. While two mechanisms, diffusion-controlled α-relaxation and surface current-related β-relaxation, have been discussed to understand the physical property of the counterion migration within the double layer in a nonuniform electric field, previous experiments on colloidal ensembles weren't able to decide the competition hence the dominance of the two. In order to solve the debate concerning the two microscopic theories, we investigate systematically the relaxation process of the electric double layer, as a function of particles size, temperature and solvent viscosity. Specifically, we use an optical tweezers-based single-particle dielectrophoretic (DEP) force spectroscopy and measure the dependence of the DEP crossover frequency. This practice avoids the presence of other electrokinetic mechanisms than DEP as complicating factors, which hindered the completeness of interpretations of published results. Finally we compare our experimental data with both predictions of α- and β-relaxations.

(420i) Ensemble Average Electrochemical TIRM: The Impact of Potential Distribution On Electrokinetic Forces

Reza M. Rock, Dennis C. Prieve and Paul J. Sides, Chemical Engineering, Carnegie Mellon University, Pittsburgh, PA

Rapid testing of electrocatalysts and corrosion resistant alloys accelerates investigation of promising new materials. Imaging amperometry, based on the deployment of colloidal particles as probes of the local current density, allows simultaneous electrochemical characterization of the entire composition space represented in an alloy film electrode. Previous work has shown that variations in particle-electrode distance for single particles in electric fields can be measured using total internal reflection microscopy, and translated into local current density, independent of electrical measurements. Implementation of this method to enable simultaneous measurements across non-uniform samples involves using a sparse, uniform layer of particles, which requires modification of previously existing theory and methods. Imaging individual particles for this application is infeasible at the low magnification levels needed to image an entire macroscopic (~1 square cm) sample. Mapping of electrochemical activity across the surface can be achieved nevertheless by imaging the entire electrode surface and gridding the resulting images into a mosaic of square "patch" areas 100 μm to a side, each containing 15-30 particles. The measured intensity in each patch is the sum of the scattering from all of the particles present in that patch. The intensity measured for these ensembles during electrochemical experiments can then be used to infer the current density across that patch.

Further improvements to the theory for translating scattering intensity to current density are proposed here, involving the assumed potential distribution on the electrode below the particles. This current distribution is governed by three dimensionless groups, which represent the exchange current density on the electrode, the balance between the ohmic and electrode potential drop in the cell, and the particle/wall separation distance. In the past, the expressions used for electrophoretic and electroosmotic force on the particle involved assuming either a uniform potential distribution or a uniform current distribution under the particle; these represent two limiting cases. Current work demonstrates that use of an intermediate case may be much better suited for the purposes of the imaging ammeter. Finite element modeling has been used here to probe a range of intermediate cases and explore the variable space of the dimensionless groups involved, to better understand the impact of parameters such as the current density and solution conductivity on the motion of the particles.

(435a) Bioactive Surfaces in Microfluidic Devices

Shashi Murthy, Chemical Engineering, Northeastern University, Boston, MA

Microfluidic channels with immobilized proteins and antibodies have been utilized for a wide range of applications, from clinical diagnostics to tissue engineering. This presentation will describe our work in the isolation of stem and progenitor cells for diagnostics and therapeutics from complex samples such as blood and digested tissue. A key element in our approach is the design of alginate hydrogel-based surface coatings that can be applied onto the surfaces of microfluidic channels. These hydrogels are functionalized with antibodies, enabling specific capture of target cells, but more importantly they also have the ability to release captured cells without loss of viability or function. Specific examples to be covered in the presentation will include isolation of endothelial progenitor cells for cardiovascular diagnostics and vascular tissue engineering and isolation of intestinal stem cells for intestinal organoid unit regeneration.

UNSTEADY FLOW FEATURES IN A T-SHAPED MICRO-MIXER

Chiara Galletti, Elisabetta Brunazzi, Roberto Mauri

Dipartimento di Ingegneria Chimica, Chimica Industriale e Scienza dei Materiali, Università di Pisa, I-56126 Pisa, Italy.

Abstract. A Direct Numerical Simulation of the mixing process between two water streams in a T-shaped micro-device indicates that for a certain range of Reynolds numbers the flow regime might be unsteady, with time-periodic structures. Pulsating flows strongly enhance the degree of mixing, that can be as large as twice the one that is observed in the steady engulfment regime. Then, when we increase the Reynolds numbers beyond the pulsating range, the flow is no more unsteady, with a consequent reduction of the mixing efficiency. In addition, we also observe that for our investigated geometry the unsteady, periodic behavior does not occur when the inlet flows are blunt and not fully developed.

Keywords: Flow instability, laminar flow, hydrodynamics, Direct Numerical Simulation.

1. INTRODUCTION

Effective mixing in small volumes is a crucial step in many processes, such as miniature fuel cells, molecular diagnostics and, in general, micro-reactors, where a fast homogenization of reactants is required. Typical uses of these microfluidic devices require these systems to be inexpensive and simple to operate, such as passive micro-mixers, where mixing is promoted by stretching and recombining of the flow fields without the help of any external power source [1].

One of the simplest designs of passive micro-mixers is a T shape, in which the inlets join the main channel with T-shaped branches. This type of mixer is also suitable for carrying out fundamental studies, as the T-shaped micro-mixer is often encountered as a junction element in more complex micro-systems. The efficiency of T-shaped micro-mixers for liquid mixing has been largely proved in literature [2].

So far, different flow regimes have been identified in the T-shaped micro-mixer depending on the Reynolds number, *Re*, namely: the stratified, the vortex and the engulfment flow regimes [3][4][5][6]. In the stratified flow regime, the inlet streams come into contact in the mixing zone and then flow side by side through the mixing channel, resulting in a completely segregated flow. In the vortex regime, above a critical *Re*, a secondary flow in the form of a double vortex pair occurs, due to the instabilities induced by the centrifugal forces at the confluence, leaving however the two incoming streams well segregated. In the engulfment regime, occurring at higher *Re*, fluid elements reach the opposite side of the mixing channel, thus largely increasing the degree of mixing. Moreover, for specific *Re* ranges above the engulfment, but still in the laminar regime, some kinds of instabilities have been recently observed. For example, based on numerical simulations on a T-mixer with a 2:1 aspect ratio (i.e. the ratio between width, W, and height, H, of the mixing channel), Bothe et al. [5] reported that when $Re > 240$, thereby above the onset of engulfment occurring at $Re = 146$, the flow exhibits a periodic behavior. For T-shaped micro-mixers having the same *2:1* aspect ratio, Dreher et al. [7] focused their attention on the flow behavior above the engulfment point, confirming that when $Re > 240$ the steady state simulations do not converge and transient ones should be adopted instead. According to the authors, this was due to a periodic behavior of the flow; in particular, they identified a periodic pulsating flow for $240 < Re < 400$, characterized by a Strouhal number $0.15 < St < 0.29$. Such periodic pulsations were found to disappear for larger *Re*, i.e. when $Re > 500$, where a chaotic, pseudo-turbulent motion was observed. Importantly, such instabilities do enhance the mixing process: indeed, optimum mixing is reached for $240 < Re < 700$. An experimental evidence on the occurrence of such flow instabilities was provided also by Thomas et al. [8][9], who investigated through planar and discrete-point Laser Induced Fluorescence the water flow in a T-shaped channel with $W = 2W' = 40\ mm$ and $H = 20\ mm$ for $56 < Re < 422$.

The present work shows results from the Direct Numerical Simulation of a T-shaped micro-mixer, characterized by a different aspect ratio than the one used by previous investigators [5][7][8][9], also for Reynolds numbers above the engulfment regime, discussing the features of flow instabilities and their relevance for mixing. Moreover, the effect of inlet flow conditions on the instability region is discussed, as understanding factors affecting the instabilities may help mixing optimization and control.

2. TEST CASE AND NUMERICAL MODEL

In the micro-mixer that we have considered (see Figure 1) the mixing channel has a rectangular cross section with a *3:2* aspect ratio (i.e. width $W = 3H/2$ and depth *H*), with length $L = 15\ H$, while the inlet channels are identical, with a *3:4* aspect ratio (i.e. width $W' = 3H/4$ and depth *H*) and length $L' = 7H/4$. Consequently, as we assume that the two inlet streams have equal flow rates, the mean velocity *V* at the inlets is equal to that in the mixing channel; in addition, flow rates were varied so that *Re* in the mixing channel ranges from *100* to *530*, with *Re* defined in terms of the hydraulic diameter, $D = 6H/5$.

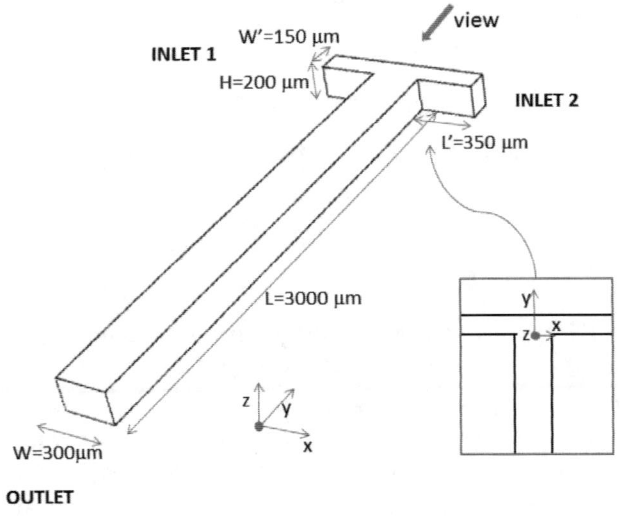

Figure 1 - Sketch of the T micro-mixer.

Simulations were performed using the commercial software Ansys 13, using the FLUENT fluid dynamics package, where length and time were scaled in terms of *D* and *D/V* [10].

After a careful dependency study on the velocity fields (Galletti *et al.* [11]), a structured grid with regular cubical elements of *H/50* size was chosen, corresponding to *3.3 M* elements. In particular, there are *76x50* elements in each cross section within the mixing channel, thus in agreement with the recommendation by Hussong *et al.* [12].

Mixing fluids are assumed to consist of the same Newtonian liquid, i.e. incompressible and with constant viscosity, with c denoting the normalized concentration (i.e. ranging from 0 to 1) of one of the two inlet streams. Constant, e.g., ambient, pressure is set at the outlet of the mixing channel, while no-slip boundary conditions are applied to the walls. Fixed mean velocity is set at both inlets, assuming that the velocity profile is fully developed. Therefore, solving the Navier-Stokes equations with no-slip boundary conditions at the walls and with constant axial pressure gradient G [13][14], we have,

$$v(y,z) = -\frac{G}{2\mu} y(Y-y) - \frac{4GY^2}{\mu\pi^3} \sum_{k\ odd} \frac{1}{k^3} \sin\left(k\pi\frac{y}{Y}\right)\left[Cosh\left(k\pi\frac{z}{Y}\right) - Tanh\left(\frac{k\pi}{2\eta}\right) Sinh\left(k\pi\frac{z}{Y}\right)\right] \quad (1)$$

where Y and Z are, respectively, the width and depth of the inlet conduit cross sections (W' and H in our case), while η is the conduit aspect ratio. The constant pressure gradient G can be derived as a function of the mean velocity V as:

$$G = -\frac{12\mu V}{Y^2}\left[1 - \frac{192}{\pi^5}\eta \sum_{k\ odd} \frac{1}{k^5} Tanh\left(\frac{k\pi}{2\eta}\right)\right]^{-1}. \quad (2)$$

A second order discretization scheme was used for all equations, as higher order solvers were found to provide negligible corrections. Some simulations were carried out with a steady solver. A sufficient number of iterations were performed in order to obtain residuals of less than 10^{-13} for all equations. The transient simulations were performed with an explicit solver, choosing a discretization time step Δt ensuring a Courant number lower than 1, i.e.

$$\frac{\Delta t}{D/V} < \frac{1}{60}\frac{V}{V_{mx}}, \quad (3)$$

where V_{mx} is the maximum fluid velocity (approximately, $V_{mx} \cong 2\text{-}3\ V$.

3. RESULTS

Figure 2 shows the tracer distribution in the mixing channel cross-section at $y = -L/3$ for different Re. Figure 2a refers to $Re = 144$ in the vortex flow regime, and indicates a poor mixing. Figure 2b and Figure 2c show the engulfment regime at $Re = 196$ and 216, respectively, characterized by non-symmetrical vortical structures resulting in enhanced mixing. Finally, Figure 2d refers to $Re = 480$: in this case, it can also be observed the presence of vortical structures, although the pattern is different from those at lower Re. For all such conditions, a solution could be found with a steady solver. In particular, the $Re = 480$ run was also performed with a transient solver, indicating steady flow. However, for intermediate Re (ranging, approximately, from 230 to 410), an unsteady behavior was observed. Figure 3 shows an example of velocity time series evaluated through a DNS simulation for $Re = 360$ at different locations in the mixing channel.

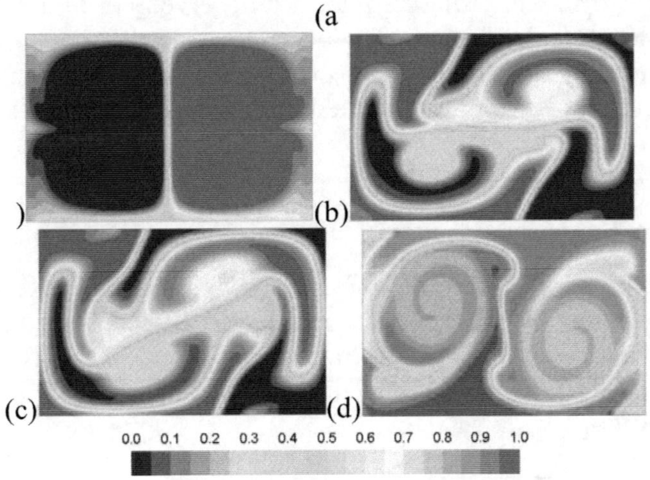

Figure 2 – Tracer distribution in the mixing channel cross section at $y = -L/3$ with (a) $Re = 144$; (b) $Re = 192$; (c) $Re = 216$ (d) $Re = 480$.

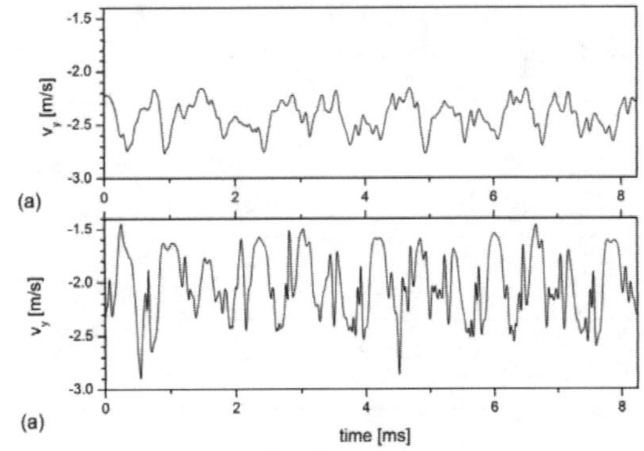

Figure 3– Time series of the y component of velocity in the center of the mixing channel for (a) $y = -2L/3$ and (b) $y = -L/3$.

In this case, it can be observed the presence of unsteady phenomena. Such time series seem to indicate a rather periodic behavior, with at least two characteristic frequencies, $f_1 \cong 0.085\ V/D$ and $f_2 \cong 0.25\ V/D$. Therefore, considering that $St = fD/V$ is the Strouhal number, we see that f_2 would lead to a $St = 0.25$, which is in excellent agreement with the result by Dreher *et al.* [7], who measured a *470 Hz* frequency using a T-mixer having $W = 2W'=2H= 600\ \mu m$ and for $Re = 300$.

Figure 4 shows the region of the T-mixer where the local velocity is above its mean value V, evaluated every $1/f$ time interval in order to describe the unsteady flow behavior. Pulsating structures originating from the interaction between the inlet streams can be observed. Consequently, the tracer distribution along the mixing channel varies greatly with time, as depicted in Figure 5 for the cross-section at $y = -L/3$. At first glance it can also be noticed that the degree of mixing is larger than that observed for all the *Re* of Figure 2, being the extension of the green region larger.

Quantitatively, the mixing efficiency is evaluated through the bulk concentration (cup mixing concentration), that is the flow-weighted mean concentration, i.e.:

$$\bar{c} = \frac{\langle cv \rangle}{\langle v \rangle}; \qquad \langle cv \rangle = \frac{1}{A}\int_A cv\, dA; \qquad \langle v \rangle = \frac{1}{A}\int_A v\, dA, \tag{3}$$

where the overbar indicates the cup mixing average, while the brackets denote the volume (i.e. un-weighted) average on a cross section of area A. The cup mixing average is preferred over the more usual volume average concentration, because it describes the concentration that "one would measure if the tube were chopped off and if the fluid issuing forth were collected in a container and thoroughly mixed" [15].

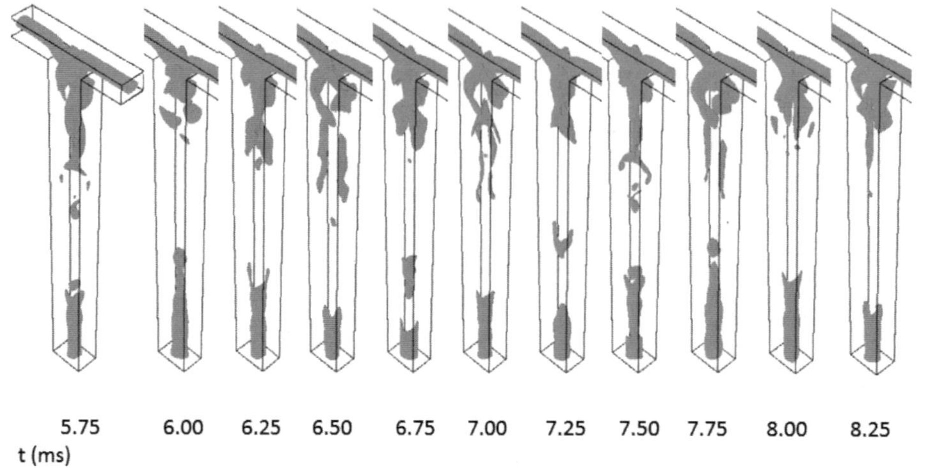

Figure 4 – Regions of the micro-mixer with velocity above *2.5 m/s*. *Re =360*.

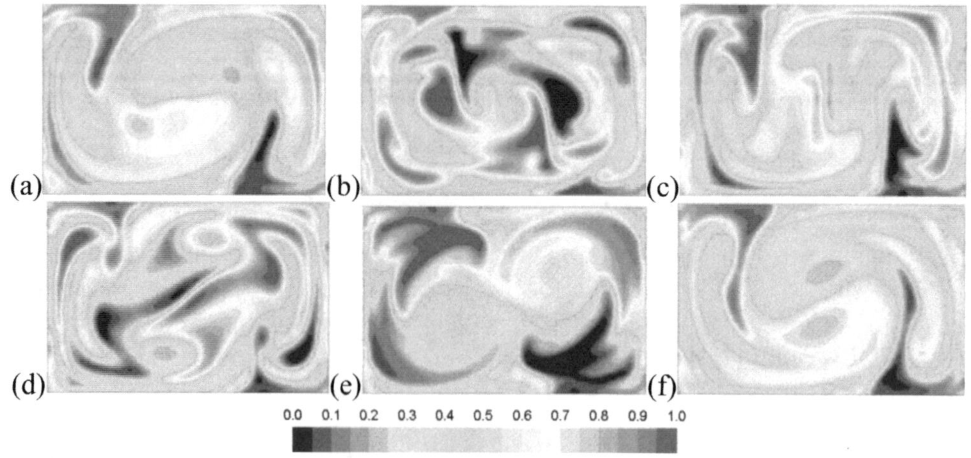

Figure 5 – Tracer distribution in the mixing channel cross-section when *Re = 360* at *y = -L/3 μm* and at different times (scaled in *D/V* unit): (a) *42.5*; (b) 44; (c) *45.5*; (d) *47 ms*; (e) *48.5 ms*; (f) *50.*

In the same way, a bulk, or cup mixing, mean square concentration difference is defined as:

$$\overline{(\Delta c)^2} = \frac{\left\langle \left(c-\bar{c}\right)^2 v \right\rangle}{\langle v \rangle}; \qquad \sigma_{cm}^2 = \overline{(\Delta c)^2} / \overline{(\Delta c)^2}_{\max}, \tag{4}$$

where its maximum value is used as a normalizing factor. Clearly, as the maximum mean square concentration difference occurs when the two streams are completely unmixed (i.e., $c = 0$ in half of the cross section and $c = 1$ in the other half) then $\overline{(\Delta c)^2}_{\max} = 1/4$. For sake of convenience, here the following definition of degree of mixing will be used:

$$\delta_m = 1 - \sigma_{cm} \tag{5}$$

as it varies from *0*, when the two streams are perfectly unmixed, to *1*, when they are completely mixed.

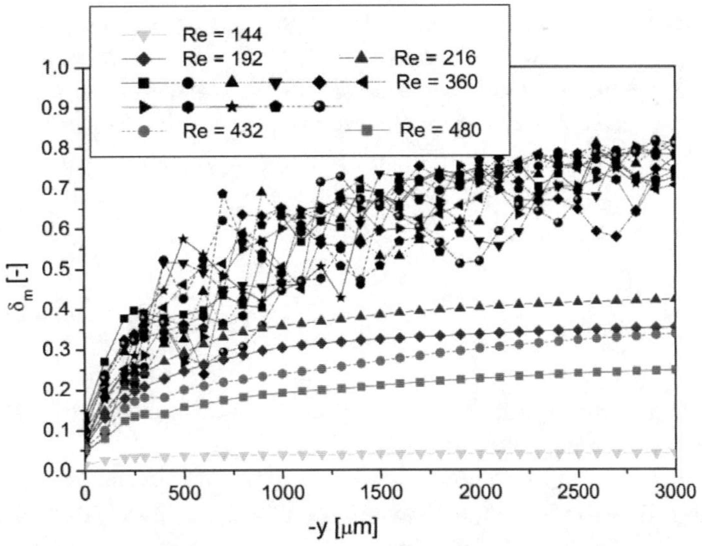

Figure 6 – Degree of mixing along the mixing channel for different regimes: vortex flow regime (green symbols); engulfment flow regime (blue symbols); unstable flow regime (black symbols); steady flow regime for high *Re* (blue symbols).

Figure 6 shows the degree of mixing along the mixing channel evaluated for different flow regimes. In the vortex regime, the degree of mixing is very low (*4%* maximum) indicating a segregated flow. In the engulfment regime, mixing is enhanced with degree of mixing at the outlet equal to *35%* when $Re = 192$, and *42%* when $Re = 216$. The unstable regime is indicated by the black symbols, each of them calculated for different time steps; the unsteady structures originate values of the degree of mixing that are well above those observed for the steady flow conditions. For instance, in the outlet section, the mixing degree varied from *70* to *82%* when $Re = 360$. Similarly also pressure drops in the T-mixer were found to vary significantly (i.e. of about *12%*) in the unstable regime for $Re = 360$. Then, as

Re is increased further, the flow field returns to be steady and the degree of mixing decreases to values that are comparable with those observed in the engulfment regime. For instance, when *Re* = *480*, we find that the in outlet section the degree of mixing is $\delta_m \cong 25\%$, which is lower than the value (i.e. $\delta_m \cong 40\%$,) at *Re* = *200*, due to the lower residence time in the mixing channel caused by the larger mean velocity.

Finally, it should be stressed that the appearing of an unsteady, pulsating regime depends not only on the geometric characteristics of the device and the operating Reynolds number, but also on the type of flow conditions at the inlets. In fact, Galletti *et al.* [11] showed that flow instabilities occur only when the velocity profile at the inlet is peaked and therefore do not appear when, for example, the inlet flow is not fully developed (i.e. the inlet channels are not long enough to ensure that fully developed flow conditions are reached).

5. CONCLUSIONS

An unsteady flow regime is observed for *Re* ranging from *228* to *408* in a T-shaped micro-mixer having a 3:2 aspect ratio, thus confirming the finding by Dreher at al. [7], who nevertheless used a 1:1 aspect ratio. The unsteady flow is characterized by pulsating structures, which strongly enhance mixing, leading to values of the degree of mixing that are almost twice as large as those achieved in the engulfment flow regime. For larger *Re*, above the unsteady flow regime, a steady flow was observed, with a strong reduction of the mixing efficiency. Therefore, the precise knowledge of the operating conditions for which the unstable flow occurs are of great importance for micro-mixer optimization. Moreover, the instabilities may be enhanced or dumped by specific conditions: for instance, Galletti *et al.* [11] showed that non-fully developed velocity profiles at the inlets may suppress all unstable flows.

REFERENCES

[1] V. Kumar, M. Paraschivoiu, K.D.P. Nigam, Single-phase fluid flow and mixing in microchannels, Chem. Eng. Sci. 66 (2011) 1329-1373.

[2] D. Bökenkamp, A. Desai, X. Yang, Y. Tai, E. Marzluff, S. Mayo, Microfabricated silicon mixers for submillisecond quench-flow analysis, Anal. Chem. 70 (1998) 232–236.

[3] M. Hoffmann, M. Schlüter, N. Räbiger, N., Experimental investigation of liquid–liquid mixing in T-shaped micro-mixers using μ-LIF and μ-PIV, Chem. Eng. Sci. 61 (2006) 2968-2976.

[4] M. Engler, N. Kockmann, T. Kiefer, P. Woias, Numerical and experimental investigations on liquid mixing in static micromixers, Chem. Eng. J. (2004) 315-322.

[5] D. Bothe, C. Stemich, H.-J. Warnecke, Fluid mixing in a T-shaped micro-mixer, Chem. Eng. Sci. 61 (2006) 2950-2958.

[6] N. Kockmann, D.M. Roberge, Transitional flow and related transport phenomena in curved microchannels, Heat Transfer Engineering 32 (2011) 595-608.

[7] S. Dreher, N. Kockmann, P. Woias, Characterization of laminar transient flow regimes and mixing in T-shaped micromixers, Heat Transfer Eng. 30 (2009) 91-100.

[8] S. Thomas, T. Ameel, J. Guilkey, Mixing kinematics of moderate Reynolds number flows in a T-channel, Phis. Fluids 22 (2009) 013601 (10 pp).

[9] S. Thomas, T. Ameel, J. Guilkey, Mixing kinematics of moderate Reynolds number flows in a T-channel, Phys. Fluids 22 (2010) 013601 (10 pp).

[10] J.M. Ottino, S. Wiggins, Introduction: mixing in microfluidics, Philos. Trans. R. Soc. London, Ser. A 362 (2004) 923-935.

[11] C. Galletti, M. Roudgar, E. Brunazzi, R. Mauri, Effect of inlet conditions on the engulfment pattern in a T-shaped micro-mixer, Chem. Eng. J. 185–186 (2012) 300-313.

[12] J. Hussong, R. Lindken, M. Pourquie, J. Westerweel, Numerical study on the flow physics of a T-shaped micro mixer, in: IUTAM Symposium on Advances in Micro- and Nanofluidics (B.V. M. Ellero et al., eds), IUTAM Bookseries 15, Springer Science, New York, 2009.

[13] P.C. Chatwin, P.J. Sullivan, The effect of aspect ratio on the longitudinal diffusivity in rectangular channels, J. Fluid Mech. 120 (1982) 347-358.

[14] J. Happel, H. Brenner, Low Reynolds Number Hydrodynamics, Prentice Hall, New York, 1965 (Eq. 2-5.24).

[15] R.B. Bird, W.E. Stewart, E.N. Lightfoot, Transport Phenomena, Wiley, New York, 1960 (pp. 297 and 641).

(435c) Stop Flow Lithography: Beyond PDMS Microfluidic Devices

Ki Wan Bong[1], Jingjing Xu[2], Jong-Ho Kim[3], Stephen C. Chapin[4], Michael S. Strano[2], Karen K. Gleason[5] and Patrick S. Doyle[4], (1)Chemical Engineering, MIT, Cambridge, MA, (2)Department of Chemical Engineering, Massachusetts Institute of Technology, Cambridge, MA, (3)Chemical Engineering, MIT, Cambridge, (4)Chemical Engineering, Massachusetts Institute of Technology, Cambridge, MA, (5)Chemical Engineering, Massachusetts Insitute of Technology, Cambridge, MA

Stop flow lithography has become a powerful technique to functionalize complex functional microparticles [1,2]. Currently, flow lithography relies on the use of polydimethylsiloxane microchannels, because the process requires local inhibition of polymerization, near channel interfaces, via oxygen permeation. The dependence on polydimethylsiloxane devices greatly limits the range of precursor materials that can be processed in flow lithography. Here we present oxygen-free flow lithography via inert fluid-lubrication layers for the synthesis of new classes of complex microparticles [3]. We use an initiated chemical vapour deposition nano-adhesive bonding technique to create non-polydimethylsiloxane-based devices. We successfully synthesize microparticles with a sub-second residence time and demonstrate on-the-fly alteration of particle height. This technique greatly expands the synthesis capabilities of flow lithography, enabling particle synthesis, using water-insoluble monomers, organic solvents, and hydrophobic functional entities such as quantum dots and single-walled carbon nanotubes. As one demonstrative application, we created near-infrared barcoded particles for real-time, label-free detection of target analytes.

1. Dendukuri, D., Gu, S.S., Pregibon, D.C., Hatton, T.A., and Doyle, P.S., "Stop flow lithography in a microfluidic device", *Lab on Chip*, **7**, 818-828, 2007.

2. Dendukuri, D., Pregibon, D.C., Collins, J., Hatton, T.A. and Doyle, P.S. "Continuous Flow Lithography for High-Throughput Microparticle Synthesis", *Nature Materials*, **5**, 365-369, 2006

3. Bong, K.W., Xu, J., Kim, J.-H., Chapin, S.C., Strano, M.S., Gleason, K.K., and Doyle, P.S., "Non-polydimethylsiloxane devices for Oxygen-Free Flow Lithography", *Nature Communications,* **3**, 805 2012.

(435d) Sculpting and Atomizing Pinned Drops with Localized Acoustic Pressures of Surface Acoustic Waves: Exponentially Small Contact Angles

Daniel Taller[1], David Go[1] and Hsueh-Chia Chang[2], (1)Aerospace and Mechanical Engineering, University of Notre Dame, Notre Dame, IN, (2)Chemical & Biomolecular Engineering, University of Notre Dame, Notre Dame, IN

Surface acoustic wave microelectromechanical (MEMS) devices are piezoelectric crystals with an interdigitated transducer on the surface of the crystal. Upon the application of a high frequency electrical signal, a surface acoustic wave (SAW) propagates along the surface of the crystal as a Rayleigh wave. Recently, SAW devices have been coupled with small amounts of liquid (~ 0.1 mm^3) in microfluidic devices. When a SAW wave interacts with a liquid film, it generates acoustic streaming enabling the manipulation of the liquid film, and, at sufficient amplitude, aerosolizes the film. Due to viscous dissipation, the SAW wave diffracting from the solid substrate into the liquid drop produces an exponentially decaying time-averaged pressure in the drop. In this work, we show that if the drop is pinned against a bounding wall such as the filter paper used in this study, the localized acoustic pressure generates a sequence of surface droplets at the contact line, whose dimensions decay in the same manner as the acoustic pressure. The undulating interfacial profile near the contact line also inherits this exponential decay, such that the averaged contact angle is exponentially small. The size distribution of surface drops is collapsed under the exponential scaling that depends only on the SAW decay rate and amplitude. This paper also presents a numerical and experimental study of the streaming and aerosolization of liquids via surface acoustic waves. A high-speed imaging system is used to characterize the fluid's free surface and aerosolization as a function of power and liquid properties. Several regimes, including a stable liquid film, initial droplet formation, thin film aerosolization, and cavitation are identified and explored. A two-dimensional numerical model based on the Laplace-Young equation is developed to model the profile of the liquid film. It is further shown that the threshold power for atomization scales as 5th power of the SAW decay rate for very localized acoustic pressures.

(435e) Microfluidic Static Droplet Arrays with Tunable Concentration Gradients

Meng Sun, Swastika S. Bithi and **Siva A. Vanapalli**, Chemical Engineering, Texas Tech University, Lubbock, TX

Recent advances in drop-based microfluidics are creating new opportunities for conducting high throughput analysis with significantly reduced sample and reagent volumes. In particular, microfluidic production of arrays of nanoliter-scale drops at prescribed coordinates could emerge as a powerful and inexpensive alternative to conventional multi-well-plate-based screening. However, the challenge is to develop effective strategies to vary the reagent concentration from drop-to-drop in the static droplet arrays (SDA). Here we describe a one-step passive strategy to create such concentration gradients in immobilized droplet arrays. The technique exploits controlled coalescence, and mixing of materials between moving plugs and stationary drops in a microfluidic parking network (MPN). We show that the concentration of one or more reagents can increase or decrease from drop-to-drop enabling a wide range of composition variation. This unique approach also generates concentration gradients of soluble reagents in drops containing insoluble materials such as particles and cells at well-defined time points.

(435f) Modeling, Predicting, and Controlling Microscale Tipstreaming

Todd Moyle[1], Lynn Walker[1] and Shelley L. Anna[2], (1)Chemical Engineering, Carnegie Mellon University, Pittsburgh, PA, (2)Departments of Mechanical Engineering and Chemical Engineering, Carnegie Mellon University, Pittsburgh, PA

Microscale tipstreaming is one method to overcome the limiting length scale in microfluidics allowing for production of submicron sized droplets. Tipstreaming is the ejection of small drops from a liquid thread formed by interfacial tension gradients and convective transport of surfactant. Controlling and understanding this process is essential for successful application in areas such as synthesis of nano-scale particles, manipulation of biomolecules, enzyme activity studies and others. However, models that predict operating conditions for tipstreaming in microfluidics do not currently exist and the periodic nature of thread formation has hindered its use in applications. In this work, we develop an analytical model aimed at capturing the essential physics of the tipstreaming mechanism. The model relies on observations of a fixed interfacial shape indicative of thread formation to simplify the fluid flow and surfactant transport equations. The result is an interfacial mass balance of surfactant. Conditions where the mass balance can be satisfied define the operating conditions for microscale tipstreaming. Good agreement is found between model predictions and experiments. Lastly, modeling the interfacial shape in this manner evokes the idea of fixing the interfacial shape experimentally. We implement an active feedback control algorithm using a custom LabVIEW code for image analysis to hold the interfacial shape fixed during tipstreaming. Controlling the interfacial shape experimentally increases thread production, potentially expanding the number of applications where tipstreaming can be used.

(435g) Block-and-Break Generation of Microdroplets with Fixed Volume

Volkert van Steijn[1], Piotr M. Korczyk[2], Adam R. Abate[3], Ladislav Derzsi[2], David A. Weitz[4] and Piotr Garstecki[2], (1)Department of Chemical Engineering, Delft University of Technology, Delft, Netherlands, (2)Institute of Physical Chemistry, Polish Academy of Sciences, Warsaw, Poland, (3)Bioengineering and Therapeutic Sciences, alifornia Institute for Quantitative Biosciences, University of California, San Fransisco, CA, (4)Department of Physics and School of Engineering and Applied Science, Harvard University, Cambridge, MA

Microfluidic devices produce monodisperse drops with a high-level of control over size and contents, making them useful for applications in biotechnology, material synthesis, optofluidics, and medical diagnostics. Drop volume depends on the flow rates, which makes it easy to tune. It, however, also sets strict requirements on the stability of flows for the generation of monodisperse drops, as variations in flow lead to corresponding variations in volume.

In this work, we introduce a novel type of microfluidic droplet generator that produces droplets of a volume set by the geometry of the droplet generator, and not by the rates of flow of the liquids. The generator consists of a classic T-junction with a bypass channel. This bypass directs the continuous fluid around the forming droplets, so that they can fill the space between the inlet of the dispersed phase and the exit of the bypass without breaking. Once filled, the droplets block the exit of the bypass and are squeezed by the continuous fluid, which leads to their pinch-off. We illustrate the potential of this fixed-volume droplet generator for i) forming monodisperse droplets from a source of varying rates of flow, ii) forming monodisperse droplets containing a gradation of solute concentration, and iii) producing monodisperse droplets in parallel.

(435h) Microbead Arrays Formed by Flow Entrapment in a Microfluidic Cell and the Effect of the Trap Geometry On the Mass Transfer of Targets to Probe Molecules On the Particle Surfaces

Xiaoxiao Chen, Chemical Engineering, The City College of New York, The School of Engineering, New York, NY and Charles Maldarelli, Levich Institute and Chemical Engineering, The City College of New York, New York, NY

Platforms which display biomolecules in an array for the multiplexed screening of the binding interactions of these molecules with molecular targets are important tools in biomedical research (drug discovery), clinical diagnostics (biomarker identification) and sensor development. Platforms which integrate the array and the binding detection in a single, microfluidic format have the advantage of reduced volumes of reagents and targets, high sensitivity, and increased throughput. Microbead-based arrays are biomolecular displays in which the probe molecules are first linked to micron-sized beads which are arranged and fixed to a surface. These arrays can easily be constructed in a microfluidic platform by using a concourse of traps designed to hydrodynamically capture the beads as they stream through the course. The arrangement of the traps serves as the template for the microbead array. By trapping the beads in a sequence of stages, in which at each stage, beads displaying only one probe are trapped, with their entrapment locations mapped, a fully indexed array can be formed. The array can then be screened against probes which are passed through the concourse.

The traps used to capture the microbeads have to properly designed to not only efficiently capture the microbeads, but also to minimize, during the assaying step, diffusion limitations in the transport of the target molecules to the surface of the captured particle. Eliminating diffusion barriers increases the dynamic response of the array, which is then only limited by the kinetics of the surface binding of the target to the probe. This presentation will examine experimentally and theoretically how trap geometries can be designed to reduce diffusion barriers. We study a general trap design consisting of an open cavity or chamber, facing the flow, in which the particle is captured within the chamber. The traps are placed in a staggered arrangement. To illustrate the effect of the trap shape on the mass transfer of the target, we examine the binding of the ligand biotin attached as the probe to the surface of the microbead, Neutravidan, which binds selectively to the biotin, is streamed through the cell. By fluorescently labeling the Neutravidin, and recording the accumulating fluorescence at each bead location by using confocal laser scanning fluorescence microscopy and epifluorescent microscopy, the rate of conjugation of the protein to the displayed ligand is measured as a function of time. Traps with large cavities sequestering beads deeply in the enclosing chamber are found to have significant diffusion barriers relative to traps in which the microbeads are held in smaller cavities. The capture efficiency is reduced by the smaller capture area of these traps, but we demonstrate that hydrodynamic arraying is still effective. Numerical solutions of the convective diffusion equations are presented for the diffusion and binding of the protein to the biotin surface of the microbeads for different trap geometries. These simulations confirm the experimental results, and provide an accurate measurement of the kinetic constants for the attachment of the protein to the biotin.

(435i) Microfluidic Hydrogel Structures for Controlled Concentration Gradients and Fast Solution Switching

Joel S. Paustian, Rodrigo Nery-Azevedo and Todd M. Squires, Chemical Engineering, University of California, Santa Barbara, Santa Barbara, CA

Hydrogels are aqueous polymer networks that are solid under stress, but are permeable to water and solutes. We present a simple microfluidic technique to synthesize in-channel hydrogel structures with precise, high resolution microscale geometries. We create create hydrogel membrane walls, with feature sizes under 10 microns, that remain sturdy under high shear from adjacent flows and withstand pressure gradients. At the same time, the hydrogel membranes allow solute, solvent, and electric fields without passing an appreciable fluid flow, thus enabling chemical conditions on one side of the membrane to be changed without mixing, stirring, or diluting.

These patterned, permeable membranes allow us to set up defined concentration or pH gradients, rapidly modify solute composition in a microchamber, or pass electric currents between channels in specified regions. Microscale dynamics, transport, and transient rheology can be probed in intriguing new ways. We provide several examples of the novel experiments enabled by these micropatterned hydrogels.

(468a) Counterion Condensation and DNA Electrophoretic Mobility

Nancy C. Stellwagen and Earle Stellwagen, University of Iowa, Iowa City, IA

The electrophoretic properties of double-stranded DNA, a relatively rigid polyanion, in free solution are very unusual because the mobility does not reflect the sizable charge on the helix backbone. Conventional wisdom posits that for each base pair added to a DNA chain, the increase in charge is offset by the increase in frictional factor so that the mobility remains constant. However, if charged groups are added to a DNA molecule without increasing its length or changing the frictional factor, the mobility remains essentially constant. This observation suggests that the counterion cloud surrounding dsDNA is modulated to maintain a constant net charge on the phosphate groups in the DNA backbone.

We have also measured the mobility of a series of 26 base pair dsDNAs containing A-tract sequences that are expected to bend the helix. Measurements were made at constant ionic strength in solutions containing various ratios of tetrabutylammonium and Li ions. A control with no A-tracts was electrophoresed at the same time, to correct for changes in the EOF and viscosity as a function of the type of cation in the solution. The bulky hydrophobic tetrabutylammonium cation preferentially increases the mobility of the A-tract DNAs, suggesting that this cation selectively attenuates the counterion cloud above A-tract sequences. The attenuation increases with increasing A-tract length but is modulated by the base composition of the flanking sequences. Systematically decreasing the tetrabutylammonium concentration and increasing the Li concentration decreases the mobility dsDNAs containing A-tracts. This decrease occurs in two discrete steps; the first step involves the return of the normal counterion cloud and the second step involves the bending of the A-tract sequence. Thus, electrophoretic measurements in the presence of variable concentrations of tetrabutylammonium cations provide a rapid and sensitive measure of A-tract length and apparent bend angle.

(468b) Testing Fundamental Theories for Polyelectrolyte Electrophoresis: Comparing Theory and Experiment As Polyelectrolyte Charge Spacing and Solvent Dielectric Constant ARE Independently Varied

David A. Hoagland, Polymer Science and Engineering, University of Massachusetts, Amherst, MA and Alexey Popov, Polymer Science and Engineering, University of Massachusetts

Standard electrostatic and electrokinetic models for polyelectrolytes depict the polymers as chains of uniform charge immersed in a continuous fluid electrolyte that dissolves point-like dissociated counterions alongside any small ions of added electrolyte. Under this depiction, the charge properties of single dilute polyelectrolytes are expressed through two dimensionless groups, a linear charge density parameter ξ and an ionic strength (or solute size) parameter $a\kappa$. The latter is the product of the polyelectrolyte's backbone radius a with the inverse of the solvent's Debye length $\kappa 1$, and the former is the ratio $e2/4\pi\varepsilon\varepsilon okTb$, where e is the electron charge, e is the solvent dielectric constant, $\varepsilon 0$ is the vacuum permittivity, k is the Boltzmann constant, T is the temperature, and b is the average contour spacing of polyelectrolyte unit charges; ξ is termed the 'Manning parameter", and the combination $e2/4\pi\varepsilon\varepsilon okT$ is termed the Bjerrum length lb, the separation between unit charges at which their mutual electrostatic interaction energy matches kT. The dimensionless mobility μo ', given by $3\eta e\mu o/2\varepsilon\varepsilon okT$, where μo is the polyelectrolyte mobility and η is the solvent viscosity, is thereby predicted by such models to be a unique function of ξ if $a\kappa$ is fixed. Classically, the counterion condensation theory of Manning provides a functional form for this ξ dependence, although other forms are available.

Taken together, expectations of the previous paragraph frame a series of fundamental tests for the assumptions that underscore standard polyelectrolyte models, with actual trends potentially explored – as done here - by electrophoresis experiments on model polyelectrolytes. To compare electrophoresis theory and experiment, it is conventional to define an effective polyelectrolyte linear charge density ξeff. This parameter captures the nonlinear electrostatic and electrokinetic impacts associated with the clustering, or condensation, of counterions along highly charged polyelectrolyte backbones. This clustering, driven by strong electrostatic attractions, reduces the polyelectrolyte's effective charge and thus its electrophoretic mobility. Clustering is absent when the polyelectrolyte linear charge is sufficiently low, and thus, $\xi eff=\xi$ for small ξ and $\xi eff<\xi$ for large ξ. Since the Debye-Hückel theory applies at low ξ, and μo ' in this limit is proportional to ξ, a measurement of μo ' at large ξ is easily transformed to a value of ξeff by extrapolating a mobility expression derived from Debye-Hückel theory.

Our key experimental strategy is to measure by capillary electrophoresis the free solution electrophoretic mobility of ionenes, the class of polyelectrolytes that contain strongly charged units (quaternized amines) in their flexible backbones. These polymers dissolve with unchanged b not just in aqueous buffers but also in other solvents of moderate to high ε; for the chosen ionenes, b itself is controlled by monomer choice, which sets the backbone chemistry between charges. Solvent mixing allows ε to be varied broadly while maintaining good solvent conditions for the polymers. Ionenes of this study dissolve in solvents such as water, ethylene glycol, acetonitrile, and n-methylformamide with $0.1<\xi<5$, the domain of greatest experimental and theoretical interest. With these beneficial properties, ionenes are uniquely suited to the proposed tests, but trends for other polyelectrolytes prove similar to those found for ionenes. Measurements for ionenes with hydrophobic spacers between charge units are indistinguishable from those for ionenes with hydrophilic spaces between these units. Further, positively charged polyelectrolytes behave no differently than negatively charged polyelectrolytes.

In high ε solvents such as water, this study's measurements uncover a linear rise of μo ' with ξ at low ξ before an asymptotic μo ' saturation as ξ exceeds approximately unity; further, as anticipated by standard theories, the measured variations of μo with b and ε collapse when expressed through ξ. Indeed, observed behaviors in high ε solvents are

reasonably consistent with theoretical predictions made long ago by Manning. In moderate ε solvents, trends are more complex. Now, at fixed aκ, μo ′ is not a unique function of ξ, and μo ′ reaches a plateau at ξ different from unity. The unexpected behaviors suggest a new controlling parameter, which we propose is the counterion size d, expressed as the dimensionless ratio lb/d (more properly,
d is regarded as the distance of closest approach between polyelectrolyte and counterion). To test our hypothesis, experiments have been repeated with counterions of different hydrated size. As the hypothesis predicts, at constant aκ, μo ′ reduces to a universal function of lb/d and ξ. The measured function surprisingly shows counterion condensation above critical values of both ξ, and lb/d, the latter effect not predicted, to our knowledge, by any theory. The unexpected condensation reflects an onset of strong ion clustering as ε deceases.

(468c) The Electrophoretic Migration of Partially Denatured dsDNA in a Gel: Why Does It Block?

Gary W. Slater, Physics, University of Ottawa, Ottawa, ON, Canada and David Sean, Physics, Université d'Ottawa, Ottawa, ON, Canada

Gel electrophoresis can separate partially denatured dsDNA fragments based on their chemical sequence. For example, upon an increase in temperature, AT-rich regions melt into two strands before GC-rich regions do. This melting changes the conformations of the DNA molecules during electrophoresis and thus affects the mobility. In practice, the mobility of partially melted DNA molecules is severely reduced, and small differences in composition often lead to large differences in mobility. In spite of the tremendous power of separation of the methods based on this well-known phenomenon, we still do not know the relation between the mobility and the conformations of the partially melted DNA molecules. The only available model, which is several decades old, predicts that the reduction in mobility is directly proportional to the average number of denatured bases regardless of their positions along the DNA chain. In this presentation, we will first derive this old model using De Gennes' reptation ideas, and we will show that the relevant parameters are not what the literature claims. We will then show how Langevin Dynamics simulations clearly demonstrate that the position of the melted regions is as important as their size, in contradiction with the assumptions made by the previous models. In fact, our results indicate that melted regions at the ends of the molecules have much more impact than melted regions (bubbles) between the two ends. Finally, again using simulations, we identify squid-like molecular conformations as being responsible for the drop in mobility upon melting. Our study also suggests new ways to optimize these methods.

(468d) DNA Gel Electrophoresis in the Entropic Trapping Regime: A Versatile Tool for Enhanced Separations and Nanostructural Analysis

Victor M. Ugaz and Nan Shi, Artie McFerrin Department of Chemical Engineering, Texas A&M University, College Station, TX

Macromolecules confined within nanoporous surroundings experience entropic trapping (ET) when their dimensions approach the average pore size, leading to emergence of transport behavior that can be immensely beneficial (e.g., a counterintuitive trend of increasing separation efficiency with DNA size during gel electrophoresis). But the noisy uncorrelated process by which the embedded macromolecules discretely hop from pore to pore contributes additional dispersion that detrimentally impacts most practical applications. Here we show how these limitations can be overcome by imposing an oscillatory electric field at a period tuned to the activation timescale of ET in ordinary hydrogels, establishing a resonance condition that synergistically combines accelerated mobility and reduced diffusion. This effect can be exploited to induce bi-directional transport of different sized DNA fragments owing to the size-dependence of the optimal modulation period. We show this in microchip electrophoresis by achieving a state in which neighboring bands in the gel travel in opposite directions.

We also show how this resonance phenomenon can be harnessed as a sensitive probe of DNA binding interactions by exploiting the fact that the conditions under which resonance emerges strongly depend on the DNA coil size. Therefore, binding interactions become visible as a distinct peak in electrophoretic mobility occurring at a conformation-dependent electric field actuation period. These measurements also enable structural information (persistence and contour lengths) to be extracted by correlating the period at maximum mobility with conformational characteristics of the DNA complex. There is no inherent lower limit on the DNA size that can be interrogated, as opposed to single-molecule studies which require large DNA (λ-phage or longer).

(468e) Rational Design of DNA Electrophoresis Devices and the Nanofence Array

Kevin D. Dorfman, Daniel W. Olson and Sung-Gyu Park, Chemical Engineering and Materials Science, University of Minnesota, Minneapolis, MN

One of the most attractive features of microfabricated separation devices is the ability to precisely define the separation medium. Given such power, the question then becomes what to fabricate. In this context, we have been working on dc electrophoretic separations of long DNA molecules in arrays of cylindrical microposts and nanoposts. Within these design constraints, the key open question is the placement of the posts. We have developed a deep understanding of the dynamics of DNA moving through arrays of obstacles, using a combination of separation experiments, single-molecule videomicroscopy, Brownian dynamics simulations and continuous-time random walk modeling. Our analysis ultimately led to the design of the so-called "nanofence" array, which features tightly spaced posts with large, open distances between them. In this presentation, I will show the logical progression that led to the design of this new type of post array, the high performance we achieved with our new device, and the surprising microscale dynamics that explain the separations in the nanofence array.

(468f) A New Type of Silicon Nanofet Detector with Single-Nanoparticle Sensitivity

Denitsa Milanova[1], Peter Griffin[2], Matthew B. Kerby[3], Thomas Niedringhaus[4], R. Fabian Pease[5] and **Annelise E. Barron**[3], (1)Mechanical Engineering, Stanford University, Stanford, CA, (2)Electrical Engineering, Stanford University, (3)Bioengineering, Stanford University, Stanford, CA, (4)Chemical Engineering, Stanford University, Stanford, CA, (5)Stanford University

The next major cost reduction in bioanalysis will be accomplished with solid-state electronic detection, eliminating the need for UV or fluorescence signals and affiliated lasers, lenses, CCDs and dyes. Jumping onto this exciting bandwagon, we have designed, fabricated, and are testing a novel type of silicon Field Effect Transistor (FET) detection system for microfluidic devices. We believe that we can integrate this detector into a low-cost platform that will be beautifully applicable to the solid-state electronic analysis of electrophoresing biomolecules from low-volume samples. We created a planar sensor—no nanopore is necessary!—and fabricated the chips using conventional, scalable CMOS techniques. These sensors can be integrated into bioanalysis devices that, in their entirety, will be the size of a USB memory stick, and which will draw their power from and download their data to a laptop computer.

Charged particles moving down a microchannel (hydrodynamically or electrophoretically) pass over a thinly insulated gate region. We took our inspiration from a new class of charged proteins (e.g., BSA with a charge of -18) altered the gate potential by ~ 1 mV, to change the source-drain current by an easily detectable ~1 nA. This sort of electronic detection technology has inherent sensitivity up to one million times greater than electrode sensing or nanopore-current blockade measurements, because it measures perturbations to the flow of electrons through the silicon chip itself. Computational methods will translate high-speed electronic pulses into quantitative signals that encode specific information about the bioanalytes of interest. This technology has myriad exciting applications in the fields of proteomics, biomarker discovery, and diagnostics by providing a label-free method to sensitively identify and quantify biological material (e.g., proteins, DNA, RNA, viruses, cells) at the single molecule/particle level. We are aiming to develop our detector for genetic analyses, and will present experimental results for the analysis of anionic nano- and microspheres; and if all goes well, for DNA molecules too.

(549a) DNA Dynamics in Nanofluidics Under Pulsed Field Electrophoresis

Wei-Ching Liao[1], Cherry Gupta[1] and L. James Lee[2], (1)Mechanical and Aerospace Engineering, The Ohio State University, Columbus, OH, (2)Chemical and Biomolecular, The Ohio State University, Columbus, OH

Nanofluidics is the study and application of fluid/particle flow in fluidic conduits at the length scale of nanometers. Due to its small size, nanofluidic devices can handle exceedingly small amount of materials, which provides high resolution for delivering, sensing, and separation of mass-limited samples. When integrated with microchannels as the inlet/outlet, the function of nanofluidic structures can be extended to the system level. Recently, a single-cell nanochannel electroporation (NEP) technique was developed in our laboratory to realize dosage-controlled delivery that could not be achieved by bulk electroporation and microchannel electroporation methods. During the operation of NEP, DNA/RNA transfection agents were electrophoretically transported into individual cells through the nanochannel under electric pulses. The gene delivery mechanism during NEP is governed by the DNA electrophoretic dynamics under pulses. In this study, we carried out Brownian dynamics simulation to investigate the electrophoretic transport of DNA through the nanochannel under electric pulses. The effect of multiple long and short pulses with the same total pulse length is studied in the simulation. Due to the recoiling of DNA chains during the interval between pulses, the transport of DNA under short pulses is not as efficient as that under long pulses. The effect of electric field distribution and DNA conformation on the degree of entrance is analyzed. Multiple-chain dynamics is also studied by simulating multiple chains with intermolecular repulsion. Due to the chain retreatment and the blocking effect, the intermolecular interaction is stronger and the reduction of transportation rate is higher under short pulses. Experimental observations on single DNA dynamics are also implemented in order to verify our simulations.

(549b) Enhanced Performance of Entropic Trap Arrays Using End-Attached Micelles

Max A. Fahrenkopf[1], Tamal Mukherjee[2], B. Erik Ydstie[1] and James W. Schneider[1], (1)Chemical Engineering, Carnegie Mellon University, Pittsburgh, PA, (2)Electrical and Computer Engineering, Carnegie Mellon University, Pittsburgh, PA

Gel-free DNA separation methods are currently designed to either separate short DNA (typically 26 to 500 bp) with high (single-base) resolution or separate long DNA (typically 5000 bp to 48500 bp) with moderate (5000 bp) resolution. Here we present a computational investigation of integrating the rapid micelle electrokinetic chromatography technique known as micelle end-labeled free solution electrophoresis (Micelle ELFSE)[1] integrated into other microfabricated devices such as an entropic trap array[2]which should serve to synergistically increase separation performance and partially bridge this gap between short and long DNA separation performance.

Micelle ELFSE utilizes end-attached non-ionic surfactant micelles that bring about a length dependent electrophoretic mobility on DNA oligomers in gel-free buffers. These micelles are tunable in size allowing one to optimize the separation performance of specific DNA target lengths. Additionally the separation device itself can be optimized with respect to geometry, such as channel dimensions and electric field strength. Entropic trap arrays are DNA separation devices that have repeating wide and narrow channel heights such that DNA is repetitively forced through a narrow constriction during electrophoresis. DNA oligomers of different lengths are separated as they cross an energy barrier upon entering confinement with a rate that depends on DNA size. Micelle ELFSE and entropic trapping are compatible separation methods since long DNA elutes first in both cases.

Here, we demonstrate that use of end-attached micelles greatly improves the performance of entropic trap arrays. Three relevant time scales were identified for DNA electrophoresis through an entropic trap that account for DNA approaching the narrow constriction, entering the constriction, and crossing the constriction[3]. We utilize Brownian dynamics simulations to investigate the sensitivity of each of these times scale to DNA tagged with a micelle. Simulations indicate a five-fold increase in resolution of the ELFSE-entropic device over simulations of the entropic trap alone. We also discuss proper choice of micelle size and electric field for greatest enhancement.

1. J.M. Savard, S.T. Grosser, J.W. Schneider *Electrophoresis* 2008, 29, 2779-2789
2. J. Han, H. Craighead *Science*, 2000, 288, 1026-1029
3. A.J. Panwar, S. Kumar *Macromolecules* 2006, 39, 1279-1289

(549c) Optimally Designed Capillary Networks for Rapid DNA Separation by Micelle End-Labeled Free Solution Electrophoresis

James W. Schneider, Max A. Fahrenkopf, Stephen Istivan, Angela Jones and B. Erik Ydstie, Chemical Engineering, Carnegie Mellon University, Pittsburgh, PA

We have recently developed a novel means to rapidly separate DNA oligomers in electric fields using end-attached surfactant micelles ("micelle-ELFSE"). The method relies on statistical fluctuations in micelle size to provide highly uniform drags for each oligomer in a population, and has long DNA elute before short DNA. Experiments verify that the method provides single-base resolution of Sanger sequencing products, up to 410 bases in length, in less than 20 minutes. An additional order-of-magnitude improvement in throughput can be realized by carefully selecting tag size, electric field, electro-osmotic flow, and capillary length for desired separation schemes. Here we present results from an optimization-based analysis of key parameters in the micelle-ELFSE process to obtain best-possible designs for micelle-ELFSE in both bench-top capillary electrophoresis (CE) and microfluidic formats.

We define a peak-dispersion model for the micelle-ELFSE system and implement the global optimization software package BARON[1], designed for non-convex, non-linear optimization problems. This optimization framework also allows us to investigate the performance of alternative capillary configurations such as parallel capillaries, microfluidic serpentines and spirals.

An important insight is that the electric field must not be so large that the oligomers fail to statistically sample an adequate number of micelle sizes during the run. This effect, which limits runtime, is more pronounced for the longer populations in the sample. Runtime can be greatly decreased by reversing elution order so that the longest DNA has the greatest residence time for sampling purposes. We present physically plausible schemes that utilize electro-osmotic counterflows to achieve this goal. Run times can be further decreased using micelles with a slight negative charge, to hasten the elution of lagging short DNA. Taken together, these refinements bring runtime for single-base resolution of 410 bases to less than 4 minutes in a bench-top CE format.

1. N.V. Sahinidis, M. Tawarmalani *Mathematical Programming* 2005, 103, 2, 225-249

(549d) Self-Assembled Nanomaterials for Capillary Electrophoresis Separations of DNA

Lisa A. Holland, Chemistry, West Virginia University, Morgantown, WV and Brandon Durney, Chemistry Department, West Virginia University, Morgantown, WV

DNA markers used for human and pathogen identification must be separated rapidly and reliably. Traditionally this is accomplished with slab gels or linear gels in capillary separations. We recently developed a new separation additive to enhance DNA sieving using phospholipids that self-assemble forming entangled ribbons. Phospholipid preparations comprised of dimyristoylphosphatidylcholine (DMPC) and dihexanoylphosphatidylcholine (DHPC) spontaneously assembly to form disks or wormlike micelles. The morphology of this nanomaterial is temperature dependent. For example, a 7 °C change in temperature can produce a ~370-fold increase in viscosity. Because the viscosity of the material is reversibly switched from a freely flowing fluid to a pseudo-gel, the sieving agent is easily loaded and expelled from microscale channels. The additive provides effective capillary separations of DNA based on chemical sieving with theoretical plate counts as high as 2,600,000. The properties and separation performance of the material are reported. Innovative separation and detection strategies for DNA are presented and the promise of a wide range of applications is discussed.

(549e) The Role of Order in DNA Separations in Colloidal Crystals

Scott B. King and Kevin D. Dorfman, Chemical Engineering and Materials Science, University of Minnesota, Minneapolis, MN

The advances in microfluidic separations of biomolecules such as DNA rely in many ways on the fabrication of the devices. Photolithographic techniques allow for detailed structures to be patterned over large areas in very repeatable locations, leading to highly ordered devices. These techniques fall under the moniker of "top-down" fabrication. If instead of photopatterning and etching one uses self-assembly, the technique is considered "bottom-up". These devices are less deterministic in their structure, and therefore not as precisely ordered. And, until now, it has been unknown whether the order of the system is important to the separation efficacy.

We have developed an analytical setup that is capable of measuring both the local order properties as well as DNA separation performance as a function of position in the device. We use self-assembled colloidal crystals as a sieving matrix, as they are capable of forming structures with either long-range order (large grain size) and short-range order (smaller grains). By carefully controlling the crystal growth rate, we can tune the location of long- and short-range order in the device. We qualitatively measure the locations of these regions using white light Bragg diffraction imaging and quantitatively measure the order using small angle laser light diffraction.

After characterizing the colloidal crystal, we then use the device to separate a known mixture of DNA. We rapidly scan the channel, taking measurements of the dyed DNA's fluoresced intensity versus position in the channel with a small time interval in between measurements. We use the large data pool to then create pseudo-finish line experiments, showing us the separation performance at arbitrary elution lengths without needing to perform the numerous separate experiments.

By combining the measurements of the order and the DNA separation, we are able to show that order does play a role in the separation power of the device. Long-range order is shown to have nearly double the separation power of short-range order colloidal crystals. This analytical setup is key in understanding the effect local properties in heterogeneous separation media have on the separation of biomolecules.

(549f) Relationship Between Frequency and Deflection Angle in the DNA Prism

Zhen Chen and Kevin D. Dorfman, Chemical Engineering and Materials Science, University of Minnesota, Minneapolis, MN

The DNA prism is a microfluidic method for separating DNA under the influence of an asymmetric pulsed electric field. In contrast to conventional pulsed field gel electrophoresis, where the DNA all migrate in the same direction, the DNA prism leads to deflection angles that are a function of the molecular weight of the DNA. As a result, the DNA prism permits the simultaneous sizing and continuous separation of large DNA. While the original DNA prism used an ordered array of microfabricated posts (Huang et al., Nature Biotech., 2004), recent experiments have implemented this strategy in colloidal crystals (Zeng et al., Angew Chem. Int. Ed., 2008, Nazemifard et al., ibid, 2010) with different pore spacings. In relatively large pores, the DNA deflection angle exhibits a maximum as a function of frequency. In very small pore spacing, however, the deflection angle increases monotonically to a plateau, which is what we might expect from predictions of the switchback mechanism (Southern et al., Nucleic Acids Res., 1987).

We will illustrate the origin of the maximum in the deflection angle as a function of frequency by simulating the DNA prism using a sparse array of posts and a Brownian dynamics algorithm. By analyzing the microscopic details of the transport process, we show that the maximum in the deflection angle arises when the electric field changes direction with a frequency that is close to the inverse of the reorientation time. Remarkably, for frequencies near the maximum deflection angle, the DNA prism also features much stronger stretching of the DNA than we observe during dc electrophoresis in the same system. Thus, although the dc electrophoresis is well described by a geometration-like mechanism, the migration in the DNA prism is much closer to classical reputation even if the pore sizes are larger than the radius of gyration of the DNA.

(549g) Multi-Layer Microfluidic Device for Human Identification Using Capillary Gel Electrophoresis for DNA Separations and a Photonic Crystal for Detection

Brandon Durney[1], Anand Kadiyala[2], Maurya Srungarapu[2], Jeremy Dawson[2] and Lisa A. Holland[3], (1)Chemistry Department, West Virginia University, Morgantown, WV, (2)Department of Computer Science and Electrical Engineering, West Virginia University, Morgantown, WV, (3)Chemistry, West Virginia University, Morgantown, WV

Two limitations of DNA analysis technology are addressed with the development of a microfluidic device with an integrated photonic crystal for fluorescent signal enhancement. This device can improve DNA measurements by providing a portable platform and by decreasing the overall time it takes to complete an analysis. Enhancement of the fluorescent signal decreases the concentration of DNA needed for reliable detection. In addition, the time required for the amplification of target DNA sequences can be reduced. A two dimensional photonic crystal fabricated on silicon using electron beam lithography can be positioned at the end of a microfluidic separation channel designed for efficient DNA separations of short tandem repeats used for human identification. An advantage of this multi-layer device design is that the photonic crystal can be detached and then reapplied to the microfluidic device. With this plug-and-play design, fluid can be controlled within the device without the crystal being present (e.g. flushing the device between runs), but added when needed for detection. Therefore, maximizing the lifetime of the photonic crystal, but rendering the separation platform disposable. This technology provides a low-cost identification device capable of on-site DNA analysis.

(549h) Lambda-DNA Dielectrophoresis in a 3D Carbon-Electrode Micro-Post Device: Theoretical and Experimental Studies

Rodrigo Martinez-Duarte[1], Fernanda Camacho-Alanis[2], Alexander Elkholy[2], Philippe Renaud[1] and Alexandra Ros[2], (1)Microsystems Laboratory, Ecole Polytechnique Fédérale de Lausanne, Lausanne, Switzerland, (2)Department of Chemistry and Biochemistry, Arizona State University, Tempe, AZ

Dielectrophoresis (DEP) is a transport mechanism where movement of particles depends of their polarizability. In general, the dielectrophoretic migration of molecules can be toward or away from regions of high electric-field gradients depending on the permittivity of the biomolecule and the medium. A typical method to create electric field gradients is introducing metallic posts along the microfluidic device. However, this approach causes electrode reactions at high applied potentials which are one of the main disadvantages of this technique. The use of insulator posts is an alternative method for DEP that generates homogenous electric field gradients over the entire depth of a microchannel. A main issue with this method is the large potentials that may be needed to achieve sufficiently high electric field gradients to manipulate small biomolecules by DEP such as DNA which polarization and its DEP response at different lengths remain unclear. Our approach is to combine advantages of metallic and insulator posts by employing carbon electrodes for DEP also known as carbon-DEP to study DNA DEP. The fabrication process of carbon electrodes gives an additional benefit in terms of materials and processing costs since they are synthesized from inexpensive photoresist, which can be patterned by optical photolithography.

Here, we analyze the DEP behavior of lambda-DNA using carbon-DEP under AC and several flow conditions both numerically and experimentally. Rows of 3D carbon electrodes made of carbonized SU-8 photoresist on fuse silica are electrically connected such that every other row has the same polarity. The experiments were performed using fluorescence microscopy under continuous flow applying AC potential. Finally, numerical simulations using a convection-diffusion model were performed.

We found that trapping and streaming behavior depends on the frequency range and hydrodynamic flow conditions. Experimental results reveal that positive DEP trapping of DNA is observed at low frequencies and small hydrodynamic flow. In contrast, streaming DEP in which DNA is concentrated according to positive DEP is enhanced at higher frequencies and higher pressure drops. We compared these experimental observations with numerical simulations considering DNA DEP as well as a pressure driven velocity component. The numerical model captured the experimentally observed streaming behavior in reasonable quantitative and excellent qualitative agreement. These findings are significant for future analytical applications to concentrate and sort biomolecules by creating streams of different biomolecules in tailored carbon-DEP devices.

(549i) A Dielectrophoretic/Electrophoretic Device for in-Situ DNA Isolation and PCR Analysis

Michael J. Heller, Nanoengineering, University of California San Diego, La Jolla, CA, Avery Sonnenberg, University of California San Diego and Raj Krishnan, Biological Dynamics

A sample to answer AC/DC dielectrophoretic/electrophoretic device has been developed that allows the isolation of cell free circulating (cfc) DNA biomarkers and bacteria/virus, with subsequent DNA processing and in-situ PCR reactions, all carried out in the same device chamber. In our initial work, an earlier generation of DEP microarray devices were used to isolate DNA from blood samples, including cell free circulating (cfc) from Chronic Lymphocytic Leukemia (CLL) and other cancer patient whole blood samples. While the DNA could be isolated from blood samples in less than twenty minutes, the PCR genotyping had to be carried out in a separate device (PCR tube). We have now developed AC/DC electrokinetic devices and procedures that allow all steps, including the PCR reactions, to be carried out in the same chamber (in-situ). This represents a major advancement in "sample to answer" diagnostics in that the isolated DNA/RNA does not have be removed or transported into a different component device. We have also been able to carry out the in-situ (in the same device chamber) isolation and preparation of DNA from bacteria for subsequent DNA sequencing, this included lyses of the cells, proteinase treatment, recapture of the extracted DNA and sequencing related tagmentation reactions, *all done in-situ*.

Introduction - AC electrokinetic methods like dielectrophoresis (DEP) are well-known techniques for achieving effective separations of cells, nanoparticles and biomolecules. Until recently, DEP technology had remained impractical for general use with complex biological samples (undiluted blood) and in high conductance solutions (>0.1 S/m). For example, very early DEP work on separating bacteria from blood (~7-9 mS/cm) required a 50-fold dilution of the blood sample before the DEP separation process could be carried out. Many other DEP separations, including cells, virus, polystyrene nanoparticles, DNA and proteins also required low-conductance conditions. More recently, some progress has been made by several groups on carrying out DEP under high conductance conditions. We for one, have developed new DEP devices and methods that allow both high molecular weight (hmw) DNA and nanoparticles to be isolated and detected in high conductance buffers (Krishnan R., Sullivan B.D., Mifflin R.L., Esener S.C., & Heller M.J., AC electrokinetic separation & detection of DNA nanoparticles in high-conductance solutions, *Electrophoresis* **29**, 1765-1774, 2008; Krishnan R. & Heller M.J., AC electrokinetic method for enhanced detection of DNA nanoparticles, *J. Biophotonics* **2**, 253-261, 2009; Krishnan R., Dehlinger D.A., Gemmen G.J., Mifflin R.L., Esener S.C., & Heller M.J., Interaction of nanoparticles at the DEP microelectrode interface, *Electrochem. Comm.* **11**, 1661-1666, 2009). Using these special microarrays with robust platinum microelectrodes over-coated with a thin hydrogel layer, DEP separations were also carried out in undiluted whole blood, where hmw-DNA and nanoparticles could be concentrated into the DEP high-field regions over the microelectrodes and the blood (red and white) cells concentrated into the DEP low-field regions between the microelectrodes. A simple fluidic wash then removes the blood cells while the DNA and nanoparticles remain concentrated in the DEP high-field regions where they can be detected by fluorescence. Hmw-DNA could be detected at 260 ng/ml, which is a detection level suitable for cell free circulating (cfc) DNA biomarker analysis. Fluorescent 40 nm nanoparticles could be detected at 9.5 x 109 particles/ml, which is a level suitable for monitoring drug delivery nanoparticles. These DEP devices were next used to isolate cell free circulating (cfc) DNA from Chronic Lymphocytic Leukemia (CLL) and other cancer patient whole blood samples. While the cfc-DNA could be isolated from blood samples in less than twenty minutes, the PCR genotyping was carried out in a separate device (tube). We now have developed AC/DC electrokinetic devices and procedures that allow all steps, including the PCR reactions, to be carried out in the same chamber (in-situ).

Results: Figure 1 (below) shows the earlier generation electric field microarray device used for the initial DEP separation work. The expanded view shows how the DEP high-field (red) regions and DEP low-field regions (green) are created when and AC electric field is applied to nine microelectrodes.

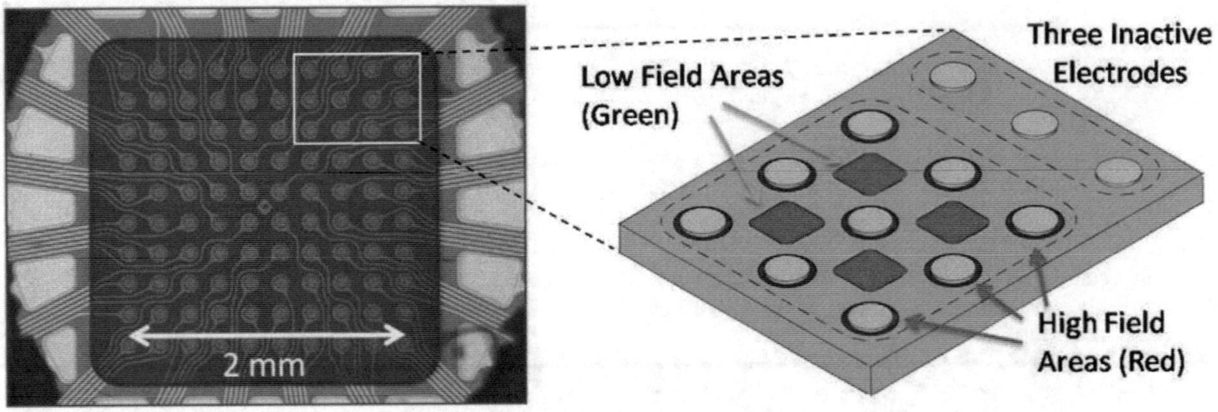

Figure 1 - DEP electric field microarray device and the DEP high-field and low-field regions

Original DC/AC electric field microarray devices were obtained from Nanogen and have one hundred 80um platinum microelectrodes overcoated with a thin hydrogel layer. Newer AC/DC dielectrophoretic/electrophoretic microarray (chip) devices specifically designed for the isolation of cfc-DNA from high conductance buffer and blood samples, as well as in-situ PCR, were obtained from Biological Dynamics. In general, DEP experiments were carried out by adding about 20 ul of blood or buffer sample to the DEP microarray device. The DEP field was applied at 10 kHz and 20 Vp-p for 15 minutes to the nine microelectrodes. The microarray was then washed three times with 0.5x PBS with the DEP field on, and finally examined by epifluorescent microscopy.

Figure 2 (below) shows the results for the DEP separation of cfc-DNA from Chronic Lymphocytic Leukemia (CLL) cancer patient whole blood samples and a normal human blood sample. All images are in green fluorescence. The DEP AC voltage was applied to the nine electrodes in columns 2, 3, 4, while no voltage was applied to the three electrodes in column 1, which serve as negative controls. Figure 2 (a) shows the results for a 20 µl sample of normal human whole blood containing 200 ng/µl SYBR green fluorescent stain, where DEP was carried out at 10 kHz and 20 Vp-p for 14 minutes, and then the microarray washed three times with 0.5x PBS. No green fluorescent rings are seen around the edges of the nine activated microelectrodes indicating no cfc-DNA is present in the normal blood sample. (Note that no fluorescence appears around the three control microelectrodes which were not activated). Figure 2 (b)-(f) shows the results for five different CLL cancer patient blood samples at different disease stages. For these CLL samples, 20 µl of undiluted whole blood containing 200 ng/µl of SYBR Green I Dye stain where added to the microarray device and DEP was carried out at 10 kHz and 20 Vp-p for 14 minutes, and then the array was washed three times with 0.5x PBS. For many of the CLL samples, green fluorescent rings can now be clearly seen around the edges of the activated microelectrodes indicating that cfc-DNA has concentrated in the DEP high-field regions. The intensity of the green fluorescent suggests that some of these CLL blood samples contain relatively large amounts of cfc-DNA. (Again note that no fluorescence appears around the three control microelectrodes.)

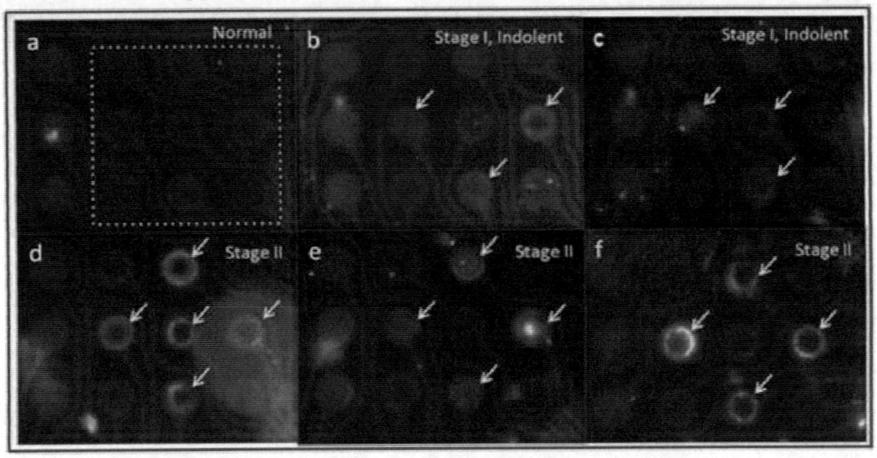

Figure 2 – (a) Normal blood after DEP (dotted white square indicates microelectrodes to which DEP was applied). (b) First CLL patient blood after DEP, yellow arrows indicate SYBR Green stained materials which are believed to be

cfc-DNA. (c) Second CLL patient blood after DEP. (d) Third CLL patient blood after DEP. (e) Fourth CLL patient blood after DEP. (f) Fifth CLL patient blood after DEP.

While cfc-DNA could be isolated from patient whole blood samples in less than twenty minutes, the cfc-DNA had to be extracted from these microarray devices and the PCR genotyping was carried out in a separate tube. We now have developed new AC/DC electrokinetic devices and procedures that allow all steps, including the PCR reactions, to be carried out in the same chamber (in-situ). Figure 3 below shows results for carrying PCR in-situ in the new AC/DC DEP/Electrophoresis microarray (chip) devices. The results show a comparison for PCR amplification of lambda phage DNA in both conventional PCR tubes and in the new AC/DC DEP microarray devices.

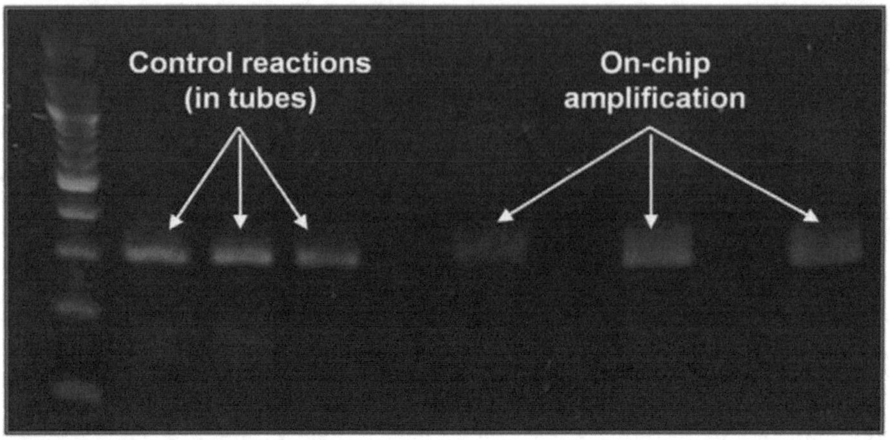

Figure 3 – PCR results for amplification of lambda phage DNA in normal PCR reaction tubes and in the new AC/DC DEP microarray (chip) devices.

Discussion and Conclusions: We have now developed AC/DC electrokinetic devices and procedures that allow all steps, including the PCR reactions, to be carried out in the same chamber (in-situ). This represents a major advancement in "sample to answer" diagnostics in that the isolated DNA/RNA does not have be removed or transported into a different component device. We have also been able to carry out the in-situ (in the same device chamber) isolation and preparation of DNA from bacteria for subsequent DNA sequencing, this included lyses of the cells, proteinase treatment, recapture of the extracted DNA and sequencing related tagmentation reactions, *all done in-situ*. The limitations for efficient DNA/RNA sample preparation have long plagued miniaturized lab on a chip devices and Point of Care (POC) diagnostic technologies, and significantly limited the overall assay sensitivity for many detection technologies. The ability to now rapidly isolated and detect disease related DNA biomarkers and bacteria/virus from undiluted whole blood will benefit many diagnostic applications by significantly reducing sample preparation time, cost and complexity.

(549j) A Novel Device for Highly Efficient Extraction of Nucleic Acids From 100 Microliter Whole Blood Samples

Lewis A. Marshall, Chemical Engineering, Stanford University, Stanford, CA and Juan G. Santiago, Stanford University, Stanford, CA

A NOVEL DEVICE FOR HIGHLY EFFICIENT EXTRACTION OF

NUCLEIC ACIDS FROM 100 MICROLITER WHOLE BLOOD SAMPLES

Lewis A. Marshall and Juan G. Santiago

Stanford University

Stanford, CA 94305, USA

We report on an effort to bridge the divide between nanoliter-scale microfluidic processes and milliliter-scale tissue sample preparation. **We have designed a novel device to electrokinetically extract nucleic acids from 100 µL of lysed whole blood with high efficiency in less than 30 minutes.** We performed this extraction using isotachophoresis (ITP), a technique that achieves selective, rapid pre-concentration and separation of molecules based on electrophoretic mobility [1]. Previous efforts to extract nucleic acids from complex biological samples using ITP have typically operated in standard etched glass microchannels or capillary systems with separation capacity equivalent to 100 nL of blood or less. As shown in Figures 1 and 2, we designed and fabricated a novel device with a separation capacity of 100 µL blood while maintaining a sample processing time of less than 30 min. **We achieve this increase in throughput in a high-aspect ratio geometry which rejects Joule heating while providing sufficient buffering capacity for electrokinetic processing.** The device was designed for a theoretical nucleic acid extraction efficiency of 100%, and we have so far demonstrated a preliminary efficiency of 25%.

The device consists of stereolithographically-defined reservoirs (Figure 2) attached to 50 x 75 mm glass layers that define the length and width of the channel. These glass layers are spaced to form a 250 µm tall separation channel with a volume of 900 µL. The device employs photolithographically fabricated phaseguides [2] to enable consistent loading of fluid into the channel without bubble formation. Each reservoir contains porous plastic spacers to hydrodynamically isolate sample and buffer chemistries. To prepare the device for operation, we loaded leading electrolyte (LE) buffer into the separation channel, and high-concentration solution into the buffering reservoirs. We pipette in 100 µL of nucleic acid solution into the sample reservoir and apply electric field to perform sample preparation. The nucleic acids are electrophoretically transferred through the channel toward the negative electrode, where they elute into the clean LE buffer. In Figure 3, we visualize operation of the device using alexa fluor 488 and fluorescein. We also demonstrated the device efficiency by processing a sample spiked with synthetic DNA oligonucliotides. We collected the output samples and analyzed them using quantitative PCR. We show recovery of 25% of dispensed synthetic DNA based on qPCR threshold cycle (as shown in Figure 4).

The device throughput is limited by heat dissipation. We deigned the device geometry to minimize temperature rise during extraction, and analyzed the theoretical steady-state temperatures. We measured steady-state temperature using thermocouples inserted into the device as a function of operational current. As shown in Figure 5, the device can operate at up to 10 mA while maintaining a temperature within 10°C of room temperature.

Flexibility in samples is a necessary component in nucleic acid sample preparation in microfluidic devices. Samples containing rare sequences are subject to Poisson statistics, so small processed volumes may not contain important targets. **To our knowledge, our system is unique in addressing separation capacity, extraction efficiency, and**

throughput in on-chip nucleic acid extraction.

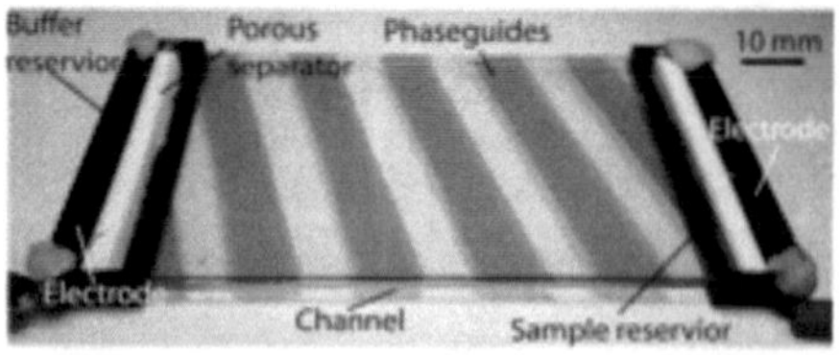

Figure 1. Image of device showing the wide central channel connecting two plastic reservoirs on each side. Blue strips are photoresist phaseguides to aid device loading. Porous spacers form a sample inlet, sample outlet, and two buffer/electrode reservoirs.

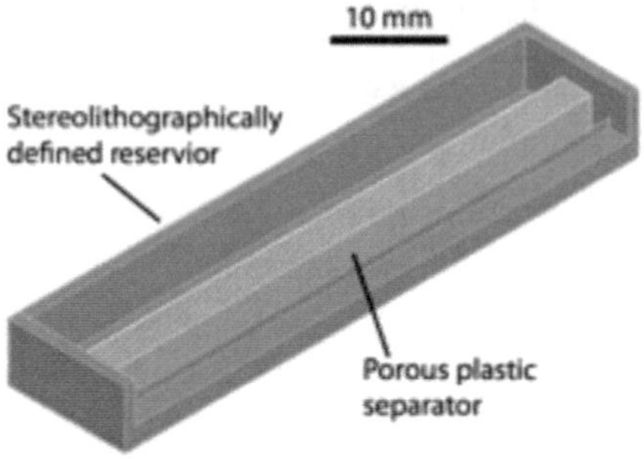

Figure 2. CAD drawing showing details of the fabricated reservoir. The device uses a stereolithographically defined plastic chamber. A porous plastic spacer separtes this into two reservoirs for electrode buffer and sample volumes. Stereolithography allows us to quickly and cheaply fabricate complex geometries for our devices.

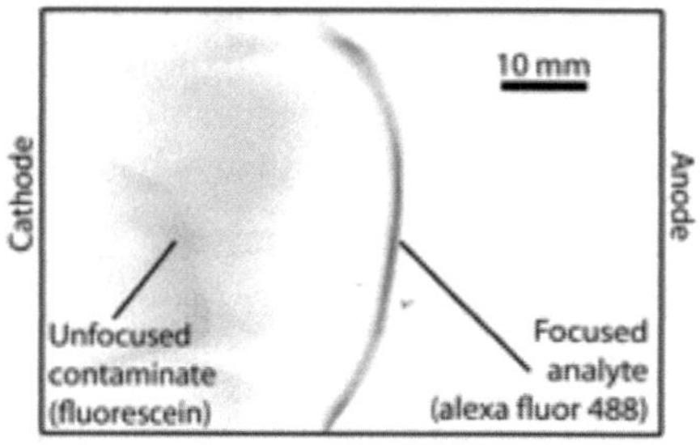

Figure 3. Fluorescence visualizaton of device operation. High-mobility Alexa Fluor 488 focuses into the ITP zone, while the lower-mobility fluorescein trails behind as an unfocused zone. The distance between the DNA simulant here (alexa fluor) increases from contaminants (fluorescein). The target sample ions (Alex Fluor) travel toward the anode, where they are collected, and the fluorescein discarded.

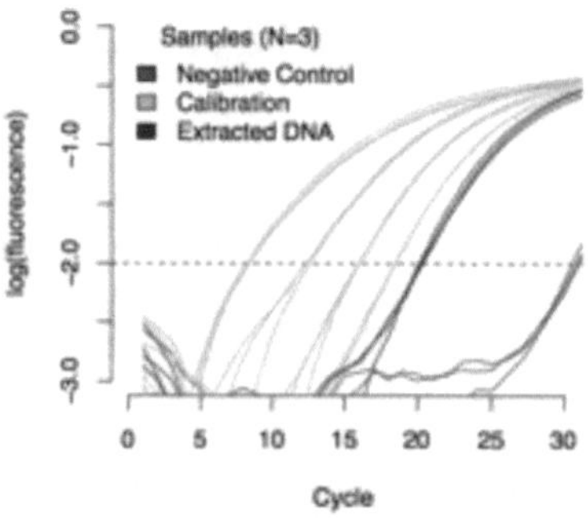

Figure 4. Off-chip quantitative PCR curves showing the purity and abundance of extracted DNA from a 100 μL sample. Based on the threshold cycle of the extracted DNA sample, we calculate a preliminary yield of 25% from the

sample. We hypothesize we can increase this to very near 100% extraction efficiency.

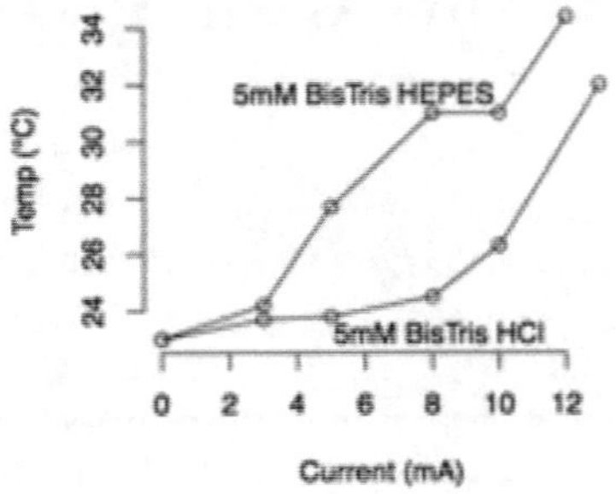

Figure 5. Measured temperature in the extraction device (maximum temperature in channel) as a function of applied current. The device throughput is limited by Joule heating, but it can operate at currents as high as 10 mA with a temperature rise of approximately 10°C. We hypothesize we can extract all DNA from 100 µl by balancing focusing dynamics, temperature, and buffering capacity.

REFERENCES:

1. "Purification of nucleic acids from whole blood using isotachophoresis," Persat, A., Marshall, L. A., and Santiago, J. G. Analytical Chemistry **81**, 9507 (2009).

2. "Phaseguides: a paradigm shift in microfluidic priming and emptying," Vulto, P., Podszun, S., Meyer, P., Hermann, C., Manz, A., and Urban, G. A. Lab Chip **11**, 1596 (2011).

(552a) Ehd Tip Streaming: Size and Charge of Electrospray Droplets

Osman A. Basaran[1], Krishnaraj Sambath[1], Robert T. Collins[2] and Michael T. Harris[1], (1)School of Chemical Engineering, Purdue University, West Lafayette, IN, (2)Oak Ridge National Laboratory, Oak Ridge, TN

When subjected to strong electric fields, pendant (sessile) drops, free drops, and thin films form conical tip structures and emit thin jets, i.e. tip stream, from these conical points that subsequently break into small droplets. This explosion of liquid drops and films subjected to an electric field into a spray consisting of charged droplets is common to mass spectrometry, printing and coating processes, and raindrops in thunderclouds. Despite being one of the oldest and most celebrated problems in science, there exist conflicting theories and measurements on the size and charge of these small electrospray droplets. While simulation can provide definitive answers to such questions, theoretical analysis of the temporal development of electrohydrodynamic (EHD) tip streaming phenomena has heretofore been elusive given the large disparity in length scales between the macroscopic drops/films and the microscopic (nanoscopic) jets. Here, simulation is used to investigate EHD tip streaming from drops and films of finite conductivity. In the simulations, the full Taylor-Melcher leaky-dielectric model, which accounts for charge relaxation, is solved to probe the mechanisms of cone formation, jet emission, and breakup of the jet into small drops. First, simulations show that tip streaming does not occur if the liquid is perfectly conducting or perfectly insulating, in accord with theory. Simulation results are used in concert with theory to develop scaling laws for the sizes of and charges carried by the micro (nano) scale drops that produced from the breakup of the thin jets. The computations at last reveal unequivocally whether electrospray droplets are Coulombically stable or unstable at the instant that they are created.

(552b) Transient Reduction of the Drag Coefficient of Charged Droplets Via the Convective Reversal of Stagnant Caps

Brad S. Hamlin and **William D. Ristenpart**, Chemical Engineering and Materials Science, University of California Davis, Davis, CA

Droplets are frequently observed to move as if they were solid rather than liquid, i.e., with no slip at the liquid-liquid interface. This behavior is usually explained in terms of the so-called "stagnant cap" model, in which surfactants accumulate at the trailing edge of the droplet, immobilizing the surface and increasing the observed drag coefficient. Here, we show that the drag coefficient for charged droplets is temporarily reduced by reversing the direction of an electric driving force. Using high speed video, we simultaneously track the velocity and relative interfacial velocity of individual aqueous droplets moving electrophoretically through oil. The observed velocity behavior is highly sensitive to the concentration of surfactant. For sufficiently low or sufficiently high concentration, upon reversal of the electric field the droplet rapidly accelerates in the opposite direction but then decelerates, concurrent with a transient rearrangement of tracer particles on the droplet surface. In contrast, droplets with intermediate surfactant concentrations exhibit neither deceleration nor significant tracer particle rearrangement. We interpret the observations in terms of convectively dominated rearrangement of the stagnant cap, and we discuss the implications for precise electrophoretic control of droplet motion in lab-on-a-chip devices and industrial electrocoalescers.

(552c) Nonlinear Electrohydrodynamics of a Viscous Droplet

Paul F. Salipante and Petia M. Vlahovska, School of Engineering, Brown University, Providence, RI

A classic result due to G.I.Taylor is that a drop placed in a uniform electric field adopts a prolate or oblate spheroidal shape, the flow and shape being axisymmetrically aligned with the applied field. We report an instability and transition to a nonaxisymmetric rotational flow in strong fields, similar to the rotation of solid dielectric spheres observed by Quincke in the 19th century. Our experiments reveal novel droplet behaviors such as tumbling, oscillations and chaotic dynamics even under creeping flow conditions. A phase diagram demonstrates the dependence of these behaviors on drop size, viscosity ratio and electric field strength. The theoretical model, which includes anisotropy in the polarization relaxation, elucidates the interplay of interface deformation and charging as the source of the rich nonlinear dynamics. Strong DC electricfields are also reported to lower the effective emulsion viscosity below that of the matrix fluid. We disuss the relation of emulsion electrorheology to drop-level dynamics.

(552d) Deformation of a Liquid Drop in a Quadrupole Electric Field

Shivraj Deshmukh and Rochish Thaokar, Department of Chemical Engineering, Indian Institute of Technology, Bombay, Mumbai, India

The two way interaction of electric field and fluid motion is a classical problem due to its importance in explaining natural phenomenon (such as thunderstorms and blue-haze found over rain forests) and many technological applications. In most of the applications electric field is seldom uniform. Moreover, in a non-uniform electric field, uncharged drops may deform and show dielectrophoretic motion which is important in many applications such as multiphase separation, biophysics and micro-fluidic systems. We study the deformation, motion and breakup of a liquid drop freely suspended in another immiscible liquid in a non-uniform axisymmetric electric field. In particular, we report the electrohydrodynamic behavior of the interface in a quadrupole electric field using analytical (asymptotic) and numerical (Boundary Element) method. The quadrupole electric field is simplest non-uniform axisymmetric electric field and when a drop is placed at the center of the field does not translate allowing systematic investigation of drop deformation and breakup. The drop behaviour is reported in a phase diagram for different electric and fluid properties. Drop in a quadrupole field shows variety of shapes depending of conductivity ratio and permittivity ratio of the fluid pair. The steady shapes are described by higher harmonics as compared to drop under uniform electric field. Positive and negative dielectrophoresis is observed when the drop is placed off center,and its translation and simultaneous deformation under quadrupole fields is also investigated. The electro-hydrostatics is unaffected by the viscosity ratio. However, the break-up of the drop and the dielectrophoretic motion and deformation strongly depend upon the viscosity ratio.

(552e) Understanding Electrocoalescence of A Two-Dimensional Arrangement of Water Droplets

Thomas F. Leary and Charles Maldarelli, Chemical Engineering, City College of New York, New York, NY

In an electrocoalescence process, conducting fluid droplets dispersed in an insulating liquid are forced to coalesce using an electric field. The process is based on the polarization of the droplets by an external electric field, which induces an attractive force required for the coalescence. Electrocoalescence has applications in a number of processes where conducting water droplets are dispersed in oil. In digital lab-on-a-chip applications, micron-sized water droplets containing different reagents and analyte samples are merged by electrocoalescence to undertake diagnostic assays. In petroleum refining, electrocoalescence has been applied for breaking emulsions in the desalting process since 1911. Despite the importance of the electrocoalescence process, a fundamental understanding of this phenomenon is still lacking, and previously reported studies focused only on one-dimensional geometries. To gain a better fundamental understanding of the process, we have studied the mechanics of electrocoalescence of a two-dimensional dispersion of water droplets in oil using a microfluidic cell. Flow focusing was used to form a monodisperse emulsion of water droplets in oil. The droplets were then directed into a microfluidic chamber whose depth was only slightly larger than the droplet diameter, causing the droplets to form a two-dimensional monolayer. A uniform electric field was applied across the monolayer and individual coalescence events were imaged using optical microscopy. Flow focusing allowed the droplet size and the emulsion number density in the chamber to be controlled. The reduced depth of the chamber allowed for the direct observation of the configuration of all the droplets, and of the merging events as a function of time.

To develop a predictive model for droplet coalescence, hydrodynamic drag coefficients were obtained by study of one-dimensional, pair-wise droplet dielectrophoresis. Finite element (FEM) numerical solutions for the quasi-static electric field around two-dimensional droplet configurations were then obtained. The sequence of observed droplet configurations, generated using video recordings, was mapped directly to the droplet location in the FEM computational domain. From these calculations, the electrostatic and hydrodynamic forces acting on all the droplets were obtained, and the force of mutual attraction or repulsion for each droplet pair was computed as a function of the droplet separation distance. These mutual forces were then correlated with the coalescence events to determine the local field strengths and pairwise droplet separation distances necessary to induce coalescence. A compilation of these individual events allowed us to construct global criteria for electrocoalescence. This model can also be used to evaluate the effect interfacial rheology, particularly the effect of surface tension on the critical electric field strength.

(552f) Electro-Hydrodynamic Effects On Lipid Membranes in Giant Vesicles

Margarita Staykova, MAE, Princeton University, Princeton, NJ, Tetsuya Yamamoto, ISIS, Strasbourg, France and Rumiana Dimova, Max Planck Institute of Colloids and Interfaces , Golm, Germany

Vesicle electro-permeabilization, -deformation and -kinetics are well documented and have found numerous practical applications [1]. Recently, using multicomponent lipid vesicles with fluorescently labeled intramembranous domains, we showed that asymmetric AC fields trigger a pronounced flow of lipids on the surface of the vesicle [2]. The flow is organized in four symmetric vortices and is fastest in the kHz frequency region. At lower and higher frequencies, it reverses direction and weakens. In this talk, we will present an analytical model according to which the membrane flow in the intermediate frequency regime is driven by the Maxwell stresses arising from the interaction between the asymmetric AC field and the induced electric charges at the vesicle interfaces. We will discuss the mechanisms of flow reversal as well as the coupling of the membrane flow to the internal and external fluid media. Finally, we will discuss the applicability of this phenomenon for fluid mixing, trapping and displacement.

1. R. Dimova et al., Soft Matter, **3**, 817 (2007)
2. M. Staykova et al., Soft matter **4**, 2168 (2008)

(552g) Autonomous Ion Current and Bubble Oscillations in a Capillary Due to Localized Film Rupture Events During Electro-Dewetting

Yu Yan, Yunshan Wang and **Hsueh-Chia Chang**, Chemical & Biomolecular Engineering, University of Notre Dame, Notre Dame, IN

Onsager reciprocal relationship stipulates that, for linear response near equilibrium, ion and mass transport driven by constant external field gradients cannot exhibit sustained oscillations in time. Yet, ion current through the ion channel of a heart cell does oscillate and so does some recent measurements of ion current through conic nanopores (see, for example, some data out of Siwy's group in JACS, 131, 5194(2009)). Such oscillations must be driven by non-equilibrium and nonlinear (possibly hysteretic) phenomena. An earlier effort (Kobatake and Fujita, J Chem Phys, 40, 2219(1964)) employs non-equilibrium mixed electro-osmotic and pressure-driven flow in a capillary, generated by two opposing concentration and pressure gradients, to produce current and flow oscillations. We show in this work that a more robust oscillation of both can be driven by an electric field, when bubble displacement by a mixed flow in a capillary is regulated by an irreversible film rupture event that results from a Maxwell interfacial instability during dewetting. We hence employ the extremely non-equilibrium, irreversible and hysteretic wetting and dewetting events to drive the observed oscillations. To sustain flow balance within and outside a nanoporous medium, an electric field through an ion-selective silica monolith within a capillary can sustain a large back pressure at the end of the monolith. We use this back pressure to compress a trapped bubble at the end of a capillary opposite from the silica monolith. At quasi-steady state, the meniscus of the bubble is stationary and the overall mixed flow rate is zero. However, the local liquid velocity is non-zero at this quasi-steady state--the system is far from thermodynamic equilibrium. The electric field is focused into the surrounding film of the trapped bubble and the corresponding ion current can be short-circuited by the film rupture. Beyond a critical voltage such that the quasi-steady state cannot be sustained, the electro-osmotic flow exceeds the pressure driven flow and expands the bubble volume by displacing its meniscus. This backward (relative to the liquid phase) meniscus motion resembles electro-dewetting except there is a liquid film surrounding the expanding bubble. Using confocal imaging, we show the film to rupture abruptly during this forward motion to shut off the field and the electro-osmotic flow. This reverses the meniscus motion and results in an opposite forward motion driven by the back pressure. The subsequent electrowetting event recoats the ruptured film to regenerate the electro-osmotic flow and restore the backward meniscus motion. It is hence the hysteretic dependence of the film thickness, due to the irreversibility of the rupture event relative to the direction of meniscus motion, and the positive feedback offered by the compressed bubble that drives this curious oscillation phenomenon. We find certain universal meniscus and current oscillation amplitude scaling that reflect the discrete rupture locations of the expanding bubble due to the Maxwell interfacial instability. The discrete rupture location offers the lag, hysteretic gap and time delay that is necessary to produce finite amplitude oscillations. We hence offer precise experimental imaging of this rupture event and correlate it to measured current/meniscus oscillations, as well as a dewetting rupture theory that includes Maxwell pressure effects. The dewetting rupture event sometimes exhibit Oswald drop coarsening pattern dynamics, reminiscent of film ruptures by molecular forces at much smaller thickness (Oron et al, Rev of Mod Phys, 69, 931(1997)).

(552h) Analytic Solutions of the Poisson-Boltzmann Equation for Nanochannels and Confined Spaces

Dimiter N. Petsev[1], Mark Fleharty[2] and Peter Crowder[2], (1)Chemical and Nuclear Engineering, The University of New Mexico, Albuquerque, NM, (2)Chemical and Nuclear Engineering, University of New Mexico, Albuquerque, NM

The electrostatic potential between charged interfaces is very important for a wide range of phenomena related to chemistry and physics of solutions. Examples include theory of electrolytes, colloid stability, and electrokinetic phenomena. The electrostatic potential in electrolyte solutions is given by the nonlinear Poisson-Boltzmann equation. This equation can be linearized for low potentials (_26 mV at room temperature.)

The case of high potentials and small separations between the charged surfaces is particularly difficult to simplify. In this paper we suggest a linearization procedure that is applicable to arbitrarily high potentials and works best in the case of small separations. This makes our approach particularly suitable for modeling nanofluidic channels. It allows us to obtain simple expressions for the potential distribution for a variety of channel shapes. We show that the analytical results can be presented in a mathematical form that is invariant for different system geometries. Figure 1 shows the mean error for the potential distribution in a cylindrical capillary. The error increases with the capillary radius, or alternatively the electrolyte concentration (inverse screening length k.) For large radii or high electrolyte concentrations,an alternative solution is derived that is based using the method of matched asymptotic expansions.

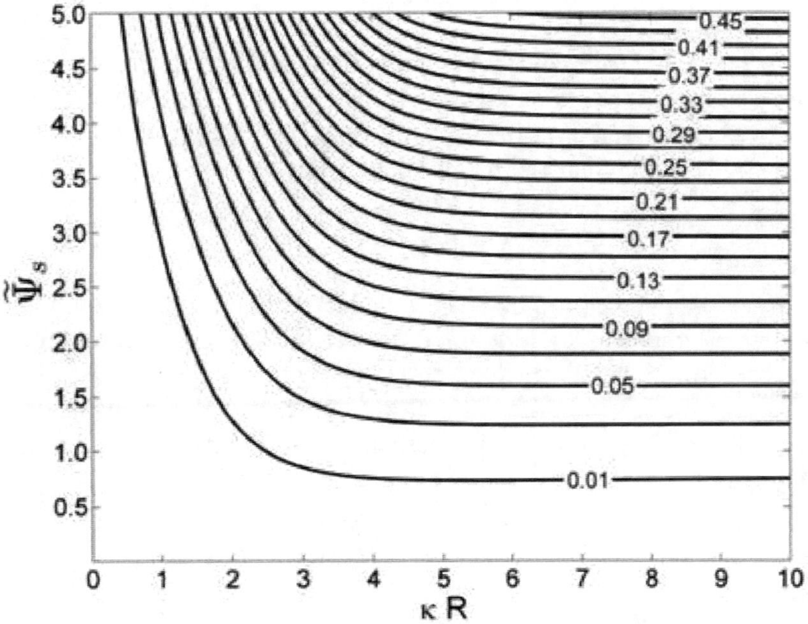

Figure 1. Contour plot of the error for the approximate solution for the electrostatic potential distribution in cylindrical channel. $\tilde{\Psi}_S = e\Psi_S / kT$ is the dimensionless surface potential. kR is the dimensionless product of the capillary radius and the inverse screening length (or inverse electric diuble layer thickness.)

Drop Dynamics Between Plate-Pin Electrode

Sameer Mhatre and Rochish Thaokar
Department of Chemical Engineering, IIT Bombay, Mumbai, India.

Introduction

An uncharged particle placed in a non-uniform electric field experiences a net force which drives the particle to regions of high or low electric fields (Pohl, 1978). This phenomena, called Dielectrophoresis (DEP), is unlike electrophoresis in which only charged particles move under any kind of field (i.e. uniform or non-uniform) in a direction that is determined by the charge on the particle. The dielectrophoresis techniques have been employed in rigid particle and bio-particle handling. Dielectrophoresis of highly deformable objects like drops and highly flexible vesicles though is not so well understood.

This paper discusses the response of a conducting drop to the non-uniformity of the applied electric field and to other aspects like field intensity, drop size, etc..

Methodology

We use pin-plate electrodes system to apply non-uniform electric field. A water drop in castor oil exhibits positive dielectrophoresis. The dynamics and the simultaneous shape change of the drop are captured using high speed camera and the videos are analysed using image analysis tool.

Results

An uncharged conductor drop undergoes dielectrophoretic motion to the pin electrode, simultaneously undergoing deformation. The conducting drop shows prolate deformation and the degree of deformation is decided by different governing parameters like field, drop size, electrical and physical properties etc.. The deformation is calculated applying Feng's theory (Feng, 1996) and compared with experimental values.

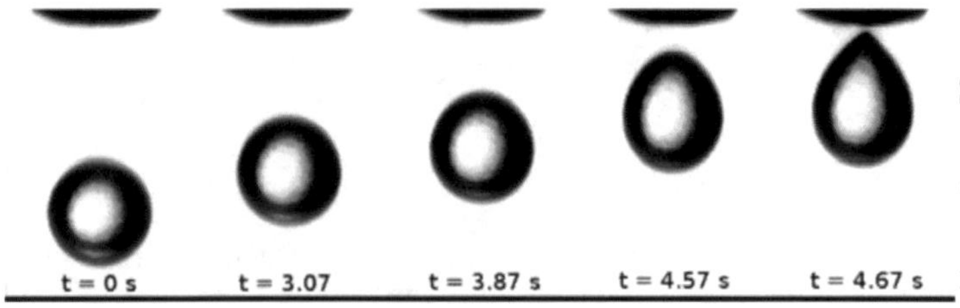

Figure 1: Snapshots of deformation and motion of a water drop in pin-plate electrodes.

Drop acquires a charge at the pin electrode and is attracted to the plate electrode. Depending upon the voltage applied, the drop can either undergo oscillations around the pin-electrode or oscillations between the two electrodes. The drop oscillation path is predicted using force balance model.

Dielectrophoretic motion is found to be in reasonable agreement with the theory.

Oscillation frequency around the pin-electrode and between the two electrodes are found to be function of applied field and drop size. A phase diagram is constructed depicting the different regimes.

The findings can be useful in applications like electric field-assisted heat and mass transfer, micro-reactors, microfluidics and in atomization techniques.

References

1. Pohl, H. A. (1978), "Dielectrophoresis the behavior of neutral matter in nonuniform electric fields", Cambridge University Press, Cambridge.

2. Feng, J. Q. (1996), "Dielectrophoresis of a deformable fluid particle in a nonuniform electric field", Phys. Rev. E, 54:4438–4441.

AUTHOR INDEX

AUTHOR INDEX

AUTHOR INDEX